SpringerBriefs in Economics

Jeffrey James

The Impact of Mobile Phones on Poverty and Inequality in Developing Countries

 Springer

Jeffrey James
Tilburg University
Tilburg
The Netherlands

ISSN 2191-5504   ISSN 2191-5512   (electronic)
SpringerBriefs in Economics
ISBN 978-3-319-27366-2   ISBN 978-3-319-27368-6   (eBook)
DOI 10.1007/978-3-319-27368-6

Library of Congress Control Number: 2015957035

Printed on acid-free paper

This Springer imprint is published by SpringerNature
The registered company is Springer International Publishing AG Switzerland

Acknowledgments

I am grateful to the secretaries in the economics department and to Ailsa Rainer for preparing the manuscript. Special thanks to Ailsa for cheerfully and accurately undertaking the vast majority of this work. It is important to express my gratitude to Research ICT Africa from Cape Town, South Africa, for making its survey data available to me.

Several of the chapters have appeared previously as journal articles under my name. Chapter 2 appeared as 'The diffusion of mobile phones in the historical context of innovations from developed countries,' Social Indicators Research, 2013 III(I), 175–184, with kind permission of Springer, Science and Business Media B.V. Chapter 6 appeared as 'Patterns of mobile phone use in developing countries: evidence from Africa,' Social Indicators Research, 2014, 119(2), 687–704, with kind permission of Springer Science and Business Media B.V. The appendix to that chapter appeared as 'Mobile phones and safety in developing countries: evidence from Sub-Saharan Africa,' GeoForum, 2015, 62, 47–50, with permission of Elsevier. Finally, Chap. 7 appeared as 'Mobile phone use in Africa: implications for inequality and the digital divide,' Social Science Computer Review, 32(1), 52–61, with permission of Sage.

I have made extensive use of World Bank datasets in the tables below. In each case I have made attribution to the Bank, name, and date of the source in question. Table 3.1 is from the World Bank, Global Economic Prospects, Technology Diffusion in the Developing World, 2008, https://openknowledge.worldbank.org/handle/10986/6334. All data are available under a CC by 3.0/GO license.

Figures 2.1 and 2.2 are based on E. Rogers, Diffusion of Innovations, 5th edn with permission of Simon and Schuster. Table 2.4 is taken from C. Antonelli, The Diffusion of Advanced Communication Technologies in Developing Countries, 1990, with permission of the OECD.

Table 3.6 is taken from 'Bottom of the Pyramid Expenditure Patterns on Mobile Services in Selected Emerging Asian Countries,' Information Technologies and International Development, 7(3), 2011, with permission of the publisher and authors. Table 3.7 is taken from ITU, Measuring the Information Society Report, 2014, with permission. Tables 3.8 and 3.9 are reproduced from GSM, Global Mobile Tax Review, 2011, with permission.

Figure 4.1 is taken from J. James and F. Stewart, 'New Products: A Discussion of the Welfare Effects of The Introduction of New Products in Developing Countries,' Oxford Economic Papers, 33(1), 1981, with permission. Table 4.3 is from H. Galperin and J. Mariscal, 'Poverty and Mobile Telephony in Latin America and the Caribbean,' DIRSI, 2007, with permission. Table 4.4 is from A. Bayes et al., 'Village Pay Phones and Poverty: Insights from a Grameen Bank Inititative in Bangladesh', 1999, Center for development Research, Bonn, with permission.

Contents

Chapter 1
Introduction

Abstract The purpose of this chapter is to state the main goals and findings of the book and to show how the chapters fit together. It is made clear at the outset, for example, that dealing with the impact of mobile phones on poverty cannot be done without first examining the extent to which the poor actually engage with this technology. For, if the engagement is minimal so too will be the direct impact on this group. It is with this recognition in mind that the first three chapters are devoted to the spread or diffusion of mobile telephones. These show, amongst other things, that with some exceptions there is quite a significant engagement of the poor with the new technology. Yet, despite this, and given the obvious importance of the issue, the literature on the impact of mobile phones on the poor is surprisingly scant. What of it there is, however, suggests the existence of a possible pro-poor bias at the level of individuals as well as countries.

Keywords Diffusion · Pro-poor bias · Information technology · Bottom of the pyramid (BoP) · Fixed-line phones · Africa

Given the obvious importance of the topic, the literature on the impact of mobile phones on poverty and inequality is somewhat disappointing.[1]

What literature there is, for example, is fragmented and drawn from 'grey' materials. It also tends to be disassociated from the technology and development literature and the economics of technical change. Perhaps most disappointingly, the impact studies tend largely to ignore the poverty component of the issue.

The purpose of this Brief is to redress these problems (where possible) by synthesizing the available literature; by making the diffusion of mobile phones a central rather than a neglected topic; by introducing new survey data to the discussion and by integrating existing studies more closely with concepts from the economics

---

[1]I deal throughout with simple, basic phones rather than the smart phones that are now used in most, if not all, developing countries. This is not because I think the latter are unimportant but rather that when one is concerned exclusively with the poor, as here, the former types of mobile phones are by far the most relevant.

© The Author(s) 2016
J. James, *The Impact of Mobile Phones on Poverty and Inequality in Developing Countries*, SpringerBriefs in Economics, DOI 10.1007/978-3-319-27368-6_1

of technical change as well as technology and development. In fact, as its point of departure, this book espouses the view that in order to understand the impact of mobile phones on poverty in developing countries, it is essential to know first how many members of this group are actually affected by these technologies. After all, in the extreme case where there is no diffusion to the poor, there cannot be any direct impact on them.

It is with this recognition in mind that two of the first three Chaps. 3 and 4 deal directly with the spread of mobile phones to those living in poverty in developing countries (defined mostly by bottom of the pyramid calculations). The purpose of these chapters is not only to show the extent to which this group engages with mobile phones, but also (to the extent possible) to explain the data that emerge from this exercise.

1.1 The Diffusion of Mobile Phones to the Poor

The third of the three chapters (Chap. 2) however, is a broad introduction to the diffusion of mobile phones in developing countries in the historical context of innovations from the developed world. In the latter region, innovations usually start with so-called 'early adopters' and then over time, in a trickle-down type process, get distributed to the lower deciles of the population (the 'late adopters').

In developing countries, by contrast, this process typically gets aborted at an early stage; indeed, new technologies rarely reach beyond 25 % of the population. Much of the reason for this has to do with the fact that innovations (with the notable exception of China) very largely originate in the industrialized countries (together with a few successful East Asian nations). Because most of these innovations are designed to fit in with host-country conditions, they are generally inappropriate to the socio-economic circumstances in the poorer countries of the world.

The rapid diffusion of mobile phones in developing countries, however, has radically altered this disappointing pattern. By 2013, for example, the number of mobile subscribers per 100 people had reached 55 even in what the World Bank defines as low-income nations. The diffusion curve thus departs substantially from the usual depiction for developing countries and comes to resemble instead that which describes the original s-shaped formulation. As a partial explanation for the mobile experience, this chapter introduces the important concept of leapfrogging, (or bypassing an older generation of technology in order to move directly to the newer version). The idea is that poorer countries with relatively little commitment to fixed-line phones, are better placed to benefit from mobiles than those that had heavily invested in this (fixed) type of communication system. Leapfrogging is a concept that will be used again in several subsequent chapters.

Chapter 3 is primarily about evidence of engagement with mobile telephony at the BoP in separate samples of African and Asian countries. In tables as telling as any others produced in this book, the results show that with some exceptions, there is quite a substantial involvement with mobile phones on the part of those at the BoP in the selected countries. Almost five years ago, for example, in Kenya the ownership of mobile phones among this group was as high as 65 %. In Ghana and Uganda the rates of ownership were just over and just under 50 % respectively. (The two poorest countries, Rwanda and Ethiopia, however, lagged well behind with ownership rates of 22 and 18 %). The Asian evidence is presented as the percentage expenditure on mobile services by each quintile at the BoP. In the lowest quintile for example percentage expenditure on mobile phones was 46 for Pakistan and 57 for the Philippines, indicating again, the intensity of demand for mobile services on the part of those at the very bottom of the pyramid.

Country differences in the samples were ascribed tentatively to variations in income levels and government policies that determine the affordability of mobile handsets and the price of calls. Much remains to be done however in this research area. Apart from better understanding differential country involvement in the use of mobile phones by the poor, there is a need to assess the welfare impact of a situation in which the purchase of these phones substitutes for other household items in the budget, such as food and clothing (a finding that has been reported several times in the literature).

Chapter 3 also raises the issue of how the spread of mobile phones may have influenced inequality given that they have been quite widely used at the BoP in numerous countries. In 2013, a study of this issue for 52 African countries confirmed that the mobile penetration rate is significant in a regression equation with the Gini coefficient as the dependent variable. This positive redistributive outcome is obviously also favourable for the poor.

Chapter 4 argues that the demonstrated appeal of mobile phones to the poor is based on a wide range of factors. These divide broadly into two groups, namely, the leapfrogging and appropriate technological characteristics of the technology on the one hand and what I refer to as a pro-poor context on the other. Of the technical characteristics of mobile phones, those that permit leapfrogging are especially important, for what they allow is a poor country to bypass fixed-line technology and move directly to the new generation of communications technology. Indeed, as shown in Chap. 5, the more limited is the percentage of fixed-lines, the greater will tend to be the gains from mobile technology. In practice, fixed-line phones tend to be least abundant in the poorest countries, where, in consequence, gains tend to be highest. Chapter 4 also considers another pro-poor set of factors, which, collectively, are referred to as 'the context'. What this broadly denotes are the many ways in which the poor inhabitants of developing countries use the mobile phone to their advantage, outside the normal working of the market mechanism, as this operates in developed countries. I am referring here for example to beeping (deliberately missed calls), sharing with family and friends and rentals of phone time in information-scarce settings. One should consider, among other things, that some pro-poor forms of use may not allow incoming calls, a topic which demands further investigation because it involves the welfare impact of mobile phones.

1.2 The Impact of Mobile Phones

Having dispelled the general notion that those at the BoP in developing countries make little or no use of mobile phones, the issue becomes one of assessing how and to what extent benefits are actually imparted to this group. In fact, the three last chapters in the book are devoted to these topics. One of them is conducted partly at the micro level, while the two others take the entire economy as the relevant unit of analysis (in which case my concern is with the poverty of nations and the inequalities between them).

Chapter 5 deals with the impact of mobile phones on the poor at all three levels of aggregation: micro, macro and the transition of successful small projects from the former to the latter level. I have already cited examples from each of these categories in the first part of the book, but have not yet given a fuller, more systematic analysis of the issues at stake. Completing such a task is the goal of this chapter. The micro-oriented literature is mainly about how mobile phones lessen or entirely remove information imperfections in the market and the way in which a better-functioning market impinges on and improves the welfare of consumers and producers. Jensen's (2007) study of the fishing industry in Kerala before and after the introduction of mobile phones, is probably the best known in this regard, showing the radical changes wrought in the relevant market by the new technology. Unfortunately, however, neither this study, nor many like it, discuss the impact of these changes specifically on the poor.

It is important because, as argued in Chap. 4, there are many reasons to expect a pro-poor impact from mobile phones, ranging from their leapfrogging and technical characteristics, to the context in which they are typically used in developing countries. And indeed what few studies there are on poverty at the micro level, clearly point in this direction. I am thinking here for example of the fact that the consumer surplus from renting phone time in the Grameen Village Pay Phone Scheme in Bangladesh is higher among the poor than the non-poor and that income growth from an expansion of mobile phones in the labour market in South Africa benefits the poor 'significantly' and 'cell phone expansion' may be described as 'pro-poor'.

The macro studies attempt to assess the simulated effect on growth of an increase in the number of phones in the countries concerned, though they mostly do not, regrettably, take the next step of estimating the impact on poverty of increased growth. This is also, in my view, a deserving topic for further research.

Three of the four studies that were reviewed found that the growth effect is larger the more scant is the percentage of fixed-line phones in a country, thus confirming the predicted pro-poor impact (although one of the studies for some reason comes to the opposite conclusion).

The final section considered three cases of successful scaling-up, namely, the Community Phone Shops in South Africa, Grameen Village Phones in Bangladesh and M-PESA mobile banking in Kenya. These were viewed against a background where successes are few and far between, not only in mobile phones and

IT, but more generally in development practice as well. Though there are many differences between the scaling-up process in the three cases, one crucial similarity stands out. That is, that in each case a forceful actor in the process was firmly committed to bringing the benefits of mobile money to the poor and disadvantaged (even if that goal at times was in conflict with narrow commercial interests in the venture as a whole).

Unfortunately though, very little of the literature on these projects has seen fit to provide estimates of how and how many of the poor have benefitted from the kiosks, village phones and mobile banking, in the countries concerned. This is plainly an area for further research, as is the better integration of the literature on scaling-up phone projects, with the extensive development writings on the topic.

At its most general however, the research task is to study the ways in which mobile phones impact on the many poor who are known to interact with them. As a step in this direction, there is a more specific need to combine the diffusion studies—especially those in Chap. 4—with the impact literature contained most especially in Chap. 5. For without knowing exactly how the poor engage with mobile phones—such as by beeping, sharing and using rental markets—the analysis of impact will be flawed. Developed country models dealing with only the improvements in market mechanisms will generally be ill-prepared for the task (not least because there are millions of poor people who do not actually operate through markets: those who produce for example only for subsistence).

Chapters 6 and 7, the two country-level studies are based on the idea that what matters more for welfare is the use rather than the access to mobile phones, a notion raised briefly in the first part of the book in connection with Sen's theory of functionings. The data analysis that forms the basis of these chapters comprises one of the first comprehensive studies of mobile use in developing countries.

The finding in Chap. 6 that the poorest countries in the sample (mostly drawn from East Africa) tend to be the most intensive users of mobile technology, in the areas of economics, health, social capital and safety, has resonance with the theory that the countries with the least and least adequate infrastructure, use this technology most intensively. When this point is taken into account, the degree of inequality between rich and poor countries will tend to be smaller than if adoption alone is used. This prediction about the causes of the digital divide is tested in Chap. 7, the final chapter. Before attending to that result, however, I shall briefly discuss the appendix to Chap. 6, which is entitled 'Mobile phones and safety: further evidence'.

The appendix is motivated by two related sets of findings. The first is from the African survey on use patterns for 11 countries in the region. This reveals that safety is the most common of the use mechanisms covered, and concludes that from a list of possible uses of mobile phones, the ability to contact others in an emergency receives the highest score across the sample as a whole. The obvious question is then: why is safety so popular a use mechanism for mobile phones in the regions concerned?

And without purporting to advance much beyond a rudimentary outline of an answer, I find that it probably has to do with the interactions between crime, poverty and inequality. Policy initiatives could do worse than study the replicability of the Peace Hut initiative in Liberia, which links the police in that country with vulnerable women.

Chapter 7 posits a negative relationship between the intensity of use of mobile phones and income per head of countries in the African sample. If this is so, the inclusion of the use variable offsets the inequality in adoption rates of mobile phones in the countries concerned. From the point of view of the digital divide, the usual ratio of mobile phones in rich versus poor countries is replaced by the ratio of mobile phones multiplied by the overall use intensity for the two groups of countries. Assuming that use is inversely related to per capita income, the digital divide will then fall. According to the survey data for 11 African countries mentioned above, this is exactly what happens though the sample size is obviously limited. If this result can be widely generalized, however, it calls into question the very notion of a global digital divide in mobile phones (based as it usually is only on technology adoption).

Reference

Jensen R (2007) The digital provide: information (technology), market performance, and welfare in the South Indian fisheries sector. QJE 122(3):879–924

Chapter 2
The Diffusion of Mobile Phones in the Historical Context of Innovations from Developed Countries

Abstract This first chapter on diffusion is a broad introduction to the topic in the historical context of innovations from the developed countries. The analytical framework comprises the well-known s-shaped curve associated with Rogers, which posits that the diffusion process goes through a number of different phases, ending up with 'late adopters' (i.e. the lowest deciles of the population). But whereas this curve well approximates what occurs in the developed countries, it is not an adequate representation of diffusion in the developing world. For there, typically, the process is aborted much earlier and innovations rarely penetrate beyond a quarter of the population (a narrow group in the urban, formal sector of the economy). It is against this background that the near diametrically opposite experience of mobile phones needs to be viewed. Even in the poorest countries, for example, the subscription rate for this technology is on average more than 50 %. This unusual experience is attributed here mainly to leapfrogging—that is, the bypassing of older generations of technology and moving straight to the new. This proposition is examined further in Chap. 4.

Keywords S-shaped curve · Leapfrogging · Diffusion · Late adopters · BoP · Pro-poor bias

The well-known s-shaped diffusion of technology curve generally works well in developed countries. But how does it perform in the very different context of developing countries? Across a wide range of new technologies imported from the developed countries it works poorly. In most cases the penetration rate fails even to reach 25 % of the population. The reason for this as I see it has to do with the concentration of innovations in the rich countries and the devotion of R&D to rich rather than poor country problems (although this problem has lessened somewhat in recent years). I redraw the s-shaped curve to reflect these facts. At the other extreme, however, are technologies such as the mobile phone, which have reached a penetration rate in some developing countries that is higher than in certain developed countries. On average, moreover, the subscription rate for this product

© The Author(s) 2016
J. James, *The Impact of Mobile Phones on Poverty and Inequality in Developing Countries*, SpringerBriefs in Economics,
DOI 10.1007/978-3-319-27368-6_2

is more than 50 % even in the poorest countries in the world. The underlying reason for this unusual experience is thought to be leapfrogging, the conditions for which are especially favourable in the case of mobile phones. Therefore there is a need to redraw the curve that explains the diffusion of most new technologies in developing countries.

Conceptually, the paper revolves around the well-known S-shaped diffusion curve of technology diffusion proposed by Everett Rogers. This curve has received wide support as a theory of how technology works in developed countries. After an initial discussion of the theory I turn to a large data-base compiled by the World Bank which, going back in history and covering multiple technologies, shows that the s-shaped curve works poorly in developing countries. There, it seems, innovations get "stuck" in their diffusion trajectories well before the majority of the population has been reached. The reason for this finding is then examined. It turns out to involve the relationship between the generation of a technology and its subsequent diffusion. In particular the heavy biases in favour of technology generation in the developed countries mean that they do not usually take into account the problems and circumstances in the poor regions of the world. In recent years however a set of new information technologies has emerged and the data for mobile phones in particular, show that their pattern of diffusion is closer to and even greater than the one found in the developed countries. I find the explanation to lie in leapfrogging and examine two such technologies from this point of view, namely digital switching and mobile phones. The circumstances required for successful leapfrogging by poor countries are most favourable in the latter technology (leapfrogging is also discussed in Chap. 4).

2.1 The Rogers Diffusion Curve: Developed Versus Developing Countries

Beginning with the assumption that the proportion of the population adopting an innovation is roughly normally distributed over time, the cumulative rate of adoption can be represented by the s-shaped curve in Fig. 2.1. This general shape, however, can take different forms. For example, low-cost innovations may exhibit a rapid take off while innovations with network effects may have faster late-stage growth.

Rogers divides the population of adopters into five groups, namely, innovators, early adopters, early majority, late majority and laggards. The size of each group is given respectively as 2.5; 13.5; 34; 34; 16 (Rogers 2003). Some authors suggest that these categories can be classified by income level, running from early adopters (high) to laggards (low) but until recently no one has attempted a systematic review of the issue.

Fig. 2.1 The s-shaped curve.
Source Based on Rogers
(2003)

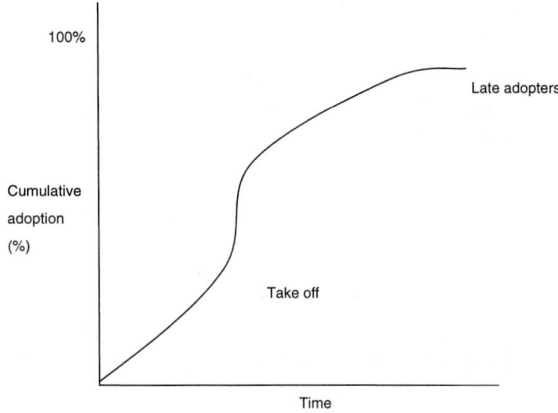

The exceptional case comprises a data-set compiled (from another source) by the World Bank (2008). In particular,

> this data set traces the extent of diffusion of some 100 technologies in 157 countries during the period 1750-2003. For each technology, only countries for which published data exist are included ... . The data analyzed here are further restricted to include only those country-technology pairs (a data set with one country and data for 7 technologies would have 7 country-technology pairs) where the intensity of use has reached at least 5 per cent of the average level of the 10 countries with the highest recorded level of diffusion. Under this restriction, there are 1,181 country-technology pairs, 699 of which correspond to developing countries (The World Bank 2008, p. 87).

One way of presenting the material thus identified is useful in comparing rates of diffusion in developed and developing countries over a number of different periods (see Table 2.1).

Table 2.1 Diffusion of innovations in rich and poor countries

	1800s			1900–1950			1950–1975			1975–2000		
	Threshold			Threshold			Threshold			Threshold		
(%)	5	25	50	5	25	50	5	25	50	5	25	50
High-income OECD	150	114	75	134	93	55	96	87	75	28	26	23
Other high income	25	16	7	28	23	14	14	10	8	7	6	6
Upper-middle	90	30	6	112	53	16	61	24	4	29	19	6
Lower-middle	109	8	2	130	38	12	33	0	0	33	5	0
Low-income	17	0	0	76	15	6	4	0	0	5	0	0
Total number of country-technology pairs World	391	168	90	480	222	103	208	121	87	102	56	35
Developing countries	216	38	8	318	106	34	98	24	4	67	24	6

Source World Bank (2008)

The main insight from the table is that diffusion in developing countries only rarely reaches a rate of 25 % and even more rarely of 50 %, with both rates well below what developed countries have achieved. 'For developing countries, the pace (and extent) of diffusion is significantly slower (lower) than in high-income countries, with only 24 (36 %) developing countries having reached the 25 % threshold and only 6 (9 %) having reached the 50 % threshold. This slower diffusion is true even for extremely old technologies, a result consistent with the idea that affordability and competency issues are binding constraints on the further diffusion of technologies in these countries' (World Bank 2008, p. 90). The corresponding figures for high-income OECD countries are that 26 out of 28 country-technologies have reached 25 and 23 % have risen to at least 50 % (see Table 2.1).

The conclusions from these and other data in the Bank report have been clearly stated by The Economist (2008) in the following terms:

> In almost all industrialised countries, once a technology is adopted it goes on to achieve mass-market scale, reaching 25 % of the market for that particular device. Usually it hits 50 %. ...

> In emerging markets this is not necessarily so. The Bank has 67 examples of a technology reaching 5 % of the market in developing countries – but only six went on to capture half the national market. Where it did catch on, it usually spread as quickly as in the West. But the more striking finding is that the spread was so rare. Developing countries have been good at getting access to technology – and much less good at putting it to widespread use.

> As a result, technology use in developing countries is highly concentrated.

Plainly, the curve drawn in Fig. 2.1 for the developed countries is not an accurate representation of what goes on in the developing world. In Fig. 2.2, I have tried to juxtapose the diffusion curves for both regions.

The main difference between the curves is that the one depicting developing countries tends to flatten out at a relatively low penetration rate.

Within developing countries it is very likely that diffusion will also be better represented over time by the non s-shaped curve in Fig. 2.2. This is not so much

Fig. 2.2 Diffusion rates over time. *Source* Based on Rogers (2003)

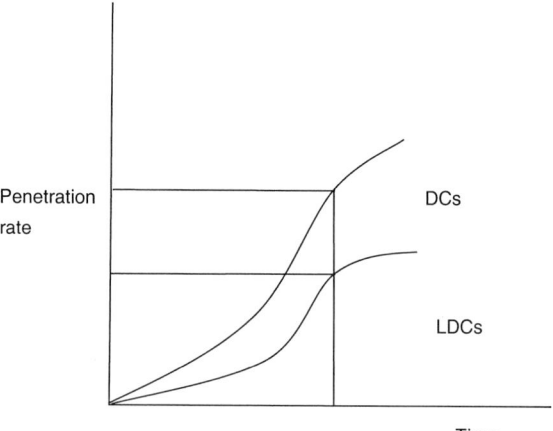

because there isn't a group of early adopters but more because rural areas are left virtually untouched by new technologies. India, for example, is one of the most technologically sophisticated developing countries with complex industries situated in and around the major cities. But advanced technologies have barely spread to the rural areas where the bulk of the less affluent members of the society is to be found. On a per capita basis, therefore, India has a relatively low level of technology diffusion. The accompanying 'skewed distribution of enterprise productivity implies potentially huge productivity and output increases are possible, if already existing within-country knowledge were to diffuse from top performers to the rest of the economy. Assuming that domestic competencies were available (or created) to efficiently use the technologies employed by enterprises at the national frontier, Indian GDP could be 4.8 times higher if those technologies were successfully applied by their less productive rivals.' (World Bank 2008, p. 92). A similar story could be told for Brazil and certain other large developing countries.

The next part of the chapter will try to show that both patterns of diffusion that have just been described between rich and poor countries and within the latter themselves—can be ascribed to the same source.

Before that, however, I wish to emphasize the implications of the patterns of technology diffusion in developing countries, for the income distribution there. For, if most technologies are confined to the relatively high-income groups (such as early adopters), the distribution of income, which is derived partly from new technologies, will tend to be highly unequal (as the following discussion of technology dualism makes clear). Certainly there is a tendency for inequality to be greater in low, rather than high-income countries, but this can be due to other factors than the diffusion of technology (such as, for example, resource abundancy and 'Dutch disease' effects).

2.2 An Explanation

With the partial exception of India and China, the vast majority of global innovative activity (measured in R&D or patents) takes place in the developed parts of the world. (especially Japan, the USA and Western Europe). According to the National Science Foundation in the United States, for example,' Although many countries conduct R and D, much of global R and D performance continues to be concentrated in a few high-income countries and regions' (NSF 2010). This is both a cause and effect of being developed: R and D expenditures helped create high incomes, while richer countries could better afford such expenditure.

The concentration of global research expenditure thus described would not necessarily constitute a bias in favour of rich countries and rich persons within countries as it is possible in principle that:

> the direction of advance, the scientific and technological priorities and the methods of solving scientific and technological problems, were independent of where the work is carried on. This, however, is patently not the case, the [then] 98 per cent of research and

development expenditures in the richer-countries are spent on solving the *problems which concern the richer countries*, according to their own priorities, and on solving these problems by the methods and approaches appropriate to the factor endowment of the richer countries. In both respects ... the interest of the poorer countries would be *bound to point in completely new directions* (Singer 1970, p. 62, emphasis added).

These directions would not however be confined, as Singer suggests, to factor endowments. For, as Stewart (1977) and others have pointed out, innovations are generated against the backdrop of a wide range of societal features. I am thinking here for example of labour skills, literacy, infrastructure and institutions. So even countries and regions with per capita incomes equal to the innovating country will tend to be lacking in these other ways. What will emerge in diffusion patterns across and within countries may be termed *technological dualism*. One part of the world (or a region within a country) has access to the innovations designed in and for the innovators while a much larger part is excluded from the benefits of research activity. The location of such activity in particular regions influences patterns of diffusion which in turn help determine the impact of new technologies.[1] There are of course exceptions to this general pattern—as when, for example, a rich country institution specifically focuses on the poor in the developing world[2]—but they represent only a small deviation from the observed tendency towards technological dualism.

The case of China, with a pronounced increase in R&D in recent years, promises a more substantial deviation. As yet however the characteristics of Chinese innovations—whether they are pro-rich or pro-poor-have not been systematically examined. But the increased capacity of this country to absorb foreign technology (even when it is complex), will certainly make its diffusion curve look more like the S-Shaped Rogers version discussed above.[3] On the demand side too, massive income increases have created the type of market for technology that is more akin to the one found in developed countries.

2.3 The Case of IT and Mobile Phones

In general and in common with other innovations from the R and D intensive countries noted above, one may expect the innovations in IT to follow the systematic patterns of diffusion and adoption that have been described as technological dualism. For one thing, one might expect the new technologies to spread most extensively to regions that closely resemble the socio-economic features of innovating countries, including per capita income as a central variable. This means the more affluent among developing countries and more advanced (urban) areas of

---

[1]A widespread diffusion for example may require a different mode of analysis compared to a more limited spread of new technology.

[2]A good example here is the One-Laptop-per-Child Programme associated with the MIT Media Lab.

[3]Though problems will occur once new technologies need to be absorbed in the rural part of the economy.

Table 2.2 Diffusion of mobile phones by country grouping (2013)

Country grouping according to World Bank	Mobile subscribers (per 100 people)
High-income	121
Upper-middle income	100
Low-middle income	85
Low-income	55

Source World Bank indicators (2014)

those parts of the world. Again though, it has to be stressed that these are tendencies rather than iron-laws. Validity may depend among other things on the type of IT with which we happen to be concerned (see below).

Quite a lot of empirical research has already gone into testing these expectations. Table 2.2 for example shows a broad correlation between income per country grouping and adoption of mobile phones.

More precise techniques (such as multiple regression analysis) have statistically confirmed the role of income and other predicted variables such as education and infrastructure (Dewan et al. 2004). Such cross-country research has sought in other words to explain the observed global digital divide. Within—country divides have also been confirmed though to a lesser extent (Donner 2008). These divides are also explained by technological dualism in that those in urban areas live in conditions that are closer to those in developed countries, from the point of view of income, skills, infrastructure, literacy and so on.

2.3.1 Exceptions: Yet Another Curve

Recall that according to World Bank research developed country innovations rarely spread beyond 25 % of the population in developing countries, a fact which I ascribed to technological dualism across and within countries. Some forms of IT however constitute a striking exception to this general pattern, especially the spread of mobile phones in developing regions. Consider the entries in Table 2.3 which show mobile penetration and per capita income for a selected sample of rich and poor countries.

The data portray a unique situation in the history of (dependent) technological relationships between rich and poor countries, namely, that some of the latter achieved (in 2010) penetration rates that are substantially higher than the former. This is the case for example with Surinam and Vietnam as against Japan and the USA. Other developing countries have reached the same rate as the two rich countries. What makes this performance all the more remarkable is that the diffusion of mobiles began later in the poor as against the rich countries.

What are the implications of these patterns for the Rogers curve? Two possibilities suggest themselves. In Fig. 2.3a I have drawn a curve for the developing countries which in the early years lies below that of the original curve for developed

Table 2.3 Rates of penetration and per capita income, selected countries, 2010

Selected developing countries 2010		Per capita income ($) 2010
Mauritius	91.67	14,000
Mongolia	91.09	3600
Nepal	109.6	1200
Mauritania	79.34	34,000
Moldova	88.59	2500
Surinam	169.64	9700
Vietnam	175.30	3100
Selected developed countries 2010		Per capita income ($) 2010
Japan	95.39	34,000
USA	89.86	47,200

Source ICT-eye, CIA World Factbook

countries (because of an initially later beginning of the process for the former). By 2010 however the rate of penetration exceeded that of the developed countries. The second possibility is where the penetration rates were roughly equal at the end of the period (though again in the early stages of the period the poor countries lag behind the rich).

I turn next to an explanation of these figures, an explanation which has much to do with technological leapfrogging.

2.3.2 Leapfrogging

The notion of leapfrogging refers to the possibility of a developing country by-passing the stages in the process through which countries were previously expected to pass. Leapfrogging generally allows diffusion to proceed more rapidly than it otherwise would. It is thus a potential explanation of the diagrams in Fig. 2.3.

Mobile phones are most readily associated with this concept but the transition from analog to digital switching technology in the 1980s also falls under the heading of leapfrogging. Digital switching allows of leapfrogging for two main reasons. One is that 'integrating the new electronic switching exchanges into an electromechanical infrastructure is much more expensive and technically complex than building a network of entirely electronic switching technology from scratch' (Antonelli 1990). The second reason is that it is developed rather than developing countries that tend to suffer from the burden of large, well-established electromechanical networks. Taken together these factors imply that developing countries 'had a remarkable opportunity to completely leapfrog the electromechanical technology, avoiding the expense of replacing obsolete (though young in age) capital stock and problems of technological cumulativity, and start their telecommunications infrastructure from scratch' (Antonelli 1990).

Fig. 2.3 New S-shaped
curves

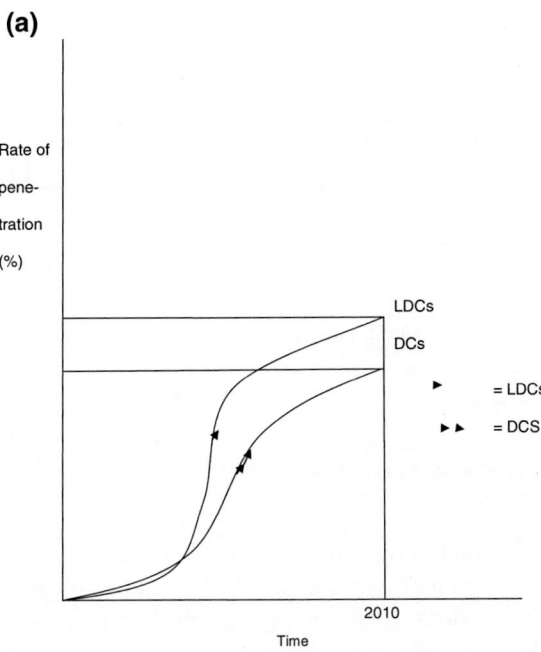

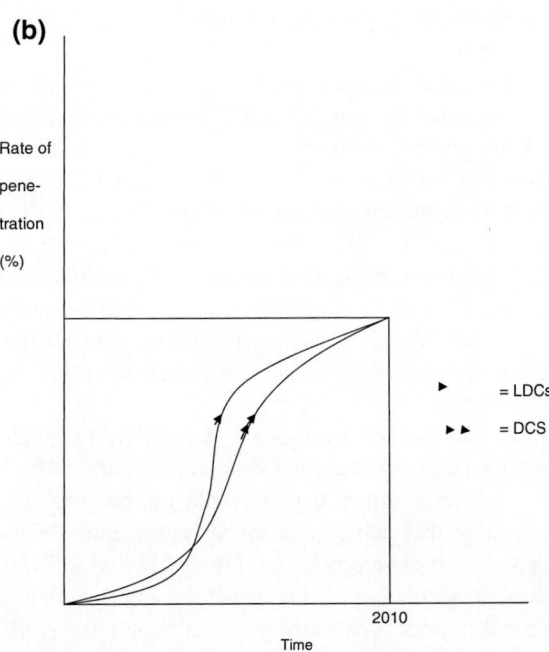

Table 2.4 Leapfrogging in telecommunications

Share of electronic switching capacity to total exchange lines (%)		Share of electronic switching capacity to total exchange lines (%)
1977		1987
Selected NICS from Asia		
Thailand	0.0	50.7
Rep. Korea	0.0	70.3
Singapore	4.0	64.5
Malaysia	7.4	64.3
Hong Kong	14.3	63.5
Selected developed countries		
USA	10.2	76.2
Canada	14.1	55.8
UK	7.0	48.4

Source Antonelli (1990)

Not all developing countries, however, were well-placed to take advantage of the opportunity thus afforded. While some of them were indeed able to leapfrog the previous technology by adopting the new switching technology more rapidly than the developed countries, others were not. The most rapid rates of diffusion occurred among the first and second-tier newly industrializing countries from Asia (see Table 2.4).

The table displays the change in the ratio of electronic lines to total switching capacity for selected NICS on the one hand and a few developed countries on the other. The most impressive case is that of Korea whose ratio went from 0 % in 1977 to over 70 % in 1987 (higher than two of the developed countries). This performance (and that of other countries in the region) was due partly to high rates of growth, investment and skills and partly also to the strategic importance that was assigned to investments in telecommunications by the governments concerned (Hanna et al. 1996). Lacking these characteristics, many other (poorer) developing countries were largely bypassed by the opportunity for leapfrogging in digital switching. Indeed many such countries continue to use analog switching technology.

Mobile phones, by contrast, have offered a much more widespread opportunity for leapfrogging in developing countries (see Table 2.5).

The table shows that over the period 2000–2013 three Sub-Saharan Africa countries had achieved a much higher ratio of mobiles to total telephones. In part the difference between Tables 2.4 and 2.5 reflects cost differences between the two technologies. For example, relative to digital switching mobile phones do not require such a costly infrastructure and skills are much less of a constraint to effective use. Much of the rapid and extensive growth of the latter technology, moreover, 'reflects the process by which it has been financed. Unlike most fixed-wire telephone systems railroads and electrical grids, mobile phone technology has been introduced into most developing countries by well- funded private

Table 2.5 Leapfrogging in mobile phones

Share of mobile to total phones		Share of mobile to total phones
2000		2013
Selected African countries		
Kenya	0.28	72
Mali	0.25	32
Ghana	0.39	108
Selected developed countries		
Canada	0.2	1.7
France	0.46	1.6
USA	0.36	2.3

Source World Bank, ICT-at—a glance tables

entrepreneurs working within a relatively competitive environment. As a result, the creation of necessary infrastructure for these systems has not been held back by the government financing and other constraints that slowed the diffusion of other technologies.' (World Bank 2008, p. 75).

The final issue I will mention concerns the uniqueness of the mobile experience. If it can be copied by other technologies there is clearly more reason to be optimistic about the future than when this experience is thought to be more nearly a one-off. The Economist (2008) clearly inclines towards the latter view, arguing that 'there are some examples of leapfrog technologies that can promote development—moving straight to local, small-scale electricity generation based on solar panels or biomass, for example, rather than building a centralised power-transmission grid—but there may not be very many.' The case of digital switching technology discussed in this chapter is another exception but I cannot come up with any more.

2.4   Conclusions

In developed countries one well-known model of the diffusion of technology is the Rogers S-shaped curve. This curve generally works quite well in the context of rich countries where the vast majority of new technologies are developed. The main task of this chapter has been to consider the shape of this curve in the *developing countries*. The first part considers what occurred in the historical perspective of technologies imported from the rich countries. The result here is that most such technologies do not spread at all in the manner described by Rogers. Most tellingly, they do not typically spread beyond 25 % of the population. The difference between rich and poor countries in this respect was attributed to the concentration of innovations in and for the former as opposed to the latter i.e. to technological dualism. At the other extreme however are certain relatively recent IT technologies that perform better in some of the poorest countries than certain developed countries. I am referring here mainly but not only to mobile phones. Here again the original S-shaped curve has to be modified to reflect this exceptional performance, which was attributed mainly to the favourable leapfrogging characteristics of mobile phones. This topic is pursued further in Chap. 4.

References

Antonelli C (1990) The diffusion of advanced telecommunications in developing countries. OECD, Paris

Dewan S, Ganley D, Kraemer K (2004) Across the digital divide: a cross-country analysis of the determinants of IT. Working paper, Graduate School of Management, University of California, Irvine. Available at http://aisel.aisnet.org/cgi/viewcontent.cgi?article=1460& context=jais. Retrieved 5 February 2008. Accessed 10 May 2010

Donner J (2008) Research approaches to mobile use in the developing world: a review of the literature. Info Soc 24(3):140–159

Hanna N, Boyson S, Gunaratne S (1996) The East Asian miracle and information technology. World Bank discussion papers, Nr. 326

National Science Foundation (2010) Science and engineering indicators. Arlington, USA

Rogers E (2003) Diffusion of innovations, 5th edn. Simon and Schuster, New York

Singer H (1970) Dualism revisited: a new approach to the problems of the dual society in developing countries. J Dev Stud 7(1):60–75

Stewart F (1977) Technology and underdevelopment. Macmillan, London

The Economist (2008) Of Internet cafes and power cuts, print edn. Feb 9. Available at http://www.economist.com/node/10640716. Accessed 1 November, 2010

World Bank (2008) Global economic prospects. Washington D.C

Chapter 3
To What Extent Are the Poor Engaged with Mobile Telephony?

Abstract The question posed by the title of this chapter is a fundamental one because of the need to establish that the poor are actually engaged with mobile phones to more than merely a minor extent. Most of the evidence on this question is available for Africa and Asia and is drawn mostly from surveys of the BoP in these regions (the African survey data are especially helpful in this regard). It turns out that with some notable exceptions, mobile phones are adopted quite widely at the BoP even in notably low-income countries and even among some of those at or near the very bottom of the pyramid. Thus, there is good reason to examine not just the association between mobile phones and the poor but also the impact of the former on the latter. Country differences are ascribed, tentatively, to deviations in income levels and government policies towards prices and taxes on handsets. There is also some African evidence in this chapter that the mobile penetration rate is significant in a regression with the Gini coefficient as the dependent variable.

Keywords Handsets · Taxes · Government policy · BoP · Africa · Asia

This chapter is based on the recognition that in order to understand the impact of mobile phones on the poor in developing countries, it is essential to know first how many members of this group are actually affected by these technologies. In the extreme case, if there is no diffusion to the poor there cannot be any direct impact, although, of course, there may be indirect effects, associated, say, with the growth in employment in the industry. According to Aker and Mbiti (2010, p. 219), for example, the dominant use of prepaid mobile phones required mobile phone companies to set up 'extensive phone credit distribution networks in partnership with the formal and informal sector' in African countries.

Before marshalling specific evidence on the question posed by the title of this chapter, however, I begin with a discussion of diffusion of mobiles at the national level, partly because from information at this aggregate level it may sometimes be possible to infer hypotheses directly about the behaviour of the poor. If, for example, it happens to be the case in a low-income country that *everyone* owns or

© The Author(s) 2016
J. James, *The Impact of Mobile Phones on Poverty and Inequality in Developing Countries*, SpringerBriefs in Economics,
DOI 10.1007/978-3-319-27368-6_3

shares a mobile phone, then there is by definition no problem of exclusion from this technology by the poor. Or, if diffusion across countries is related to their relative income levels, so too might one expect patterns of adoption within countries also to vary with this factor. Certainly, there is strong econometric evidence showing that income plays a prominent role in explaining cross-country diffusion patterns (Wheeler et al. 2001; Wesolowski et al. 2012). And as shown in Tables 3.1 and 3.2 the spread of mobile phones is least in evidence in ultra low-income countries, especially from Sub-Saharan Africa. Note for example how large is the difference between diffusion rates in such countries and the rest of the world (Table 3.2). In particular, at 47 % the average diffusion rate in Africa is just over half the corresponding figure for low-middle income countries. The latter in turn is close to the 92.4 % rate of diffusion recorded for the upper-middle income group.

3.1   Connections Versus Unique Subscribers

Table 3.3 shows the number of subscribers per 100 people for a selected sample of low and low-middle income countries in Africa and Asia. What is perhaps most striking about this table is the variation of the data even among comparable countries. In Mali, for example, mobile phones have apparently reached over 100 % of the population, while in Niger the corresponding figure is only 39 %. For Mali and other low-income countries that enjoy a subscription rate of over 100 %, it might be assumed, as noted above, that all poor inhabitants own or share a mobile phone. This assumption, however, is likely to lead to a substantial overstatement of the

Table 3.1 Mobile subscriptions per 100 people by region

Region	2012
	Subscriptions per 100 people
Europe and Central Asia	108.5
Latin America	108.1
East Asia and the Pacific	88.9
South Asia	68.9
Sub-Saharan Africa	59.3
World	89.3

Source World Bank (2014)

Table 3.2 Mobile subscriptions by country income level

Income group	2012
	Subscriptions per 100 people
High-income	122.9
Upper-middle	92.4
Lower-middle	83.1
Low-income	47.2

Source World Bank (2014)

Table 3.3 Mobile subscriptions per 100 people (selected countries)

Country	2013
	Subscriptions per 100 people
Mali (low income)	129
Moldova (low middle)	106
Myanmar (low income)	13
Niger (low income)	39
Vietnam (low middle)	131
Ghana (low middle)	108
Benin (low income)	93
Kenya (low middle)	72

Source World Bank Indicators (2013)

involvement of the poor with mobiles. For, as Gillet (2014) of the GSM Association has rightly pointed out, penetration is often measured in SIM cards rather than unique subscribers. As he himself puts it, 'The fact that there will soon be as many mobile connections as people on the planet is a huge achievement …. However, …… it would be wrong to assume that almost every single person on the planet is now connected to a mobile network. In reality only around one in two people worldwide are actually subscribed to a mobile service to date' (Gillet 2014).

The problem is that,

There is an important difference between the number of mobile connections – the metric traditionally used by the industry to measure market size and penetration – and what we term unique mobile subscribers. The latter refers to a single individual that has subscribed to a mobile service and that person can hold multiple mobile connections (i.e. SIM cards).

If one individual actively uses two SIM connections that person will be counted by the industry as two mobile connections, although he or she is only one mobile subscriber. Multiple connections ownership has been distorting mobile market penetration figures for many years, as standard practice for measuring penetration has been to divide the number of registered SIM cards by the country's population (Gillet 2014).

Gillet chooses China as an example and it is an important one since it matters a great deal in absolute terms whether one uses the conventional concept of subscribers or the superior measure known as unique mobile subscribers. Thus, that country recently reached a total of 1.25 billion connections as against a population total of 1.39 billion. 'However, this does not mean that every Chinese citizen will have subscribed to a mobile service.' Indeed, the author calculates that as of the end of 2013, subscribers each carried 1.79 SIM cards on average, which means that only 630 million of them account for the total connections of 1.25 billion (i.e. less than half the population). Given the connection between income and diffusion, one may reasonably assume that the remaining 760 million are made up disproportionately by the poorest groups in Chinese society (who, more often than not, are located in rural rather than urban areas).[1]

---

[1]For a discussion of rural versus urban patterns of adoption in developing countries, see Chick et al. (2010).

3.2 Direct Evidence

If, therefore, little can be gleaned from conventional country data on mobile phone subscribers about the prevalence of *unique* subscribers among the poor, one is left with what direct evidence is available. Regrettably, it is still largely true that studies exploring mobile phone use among 'the economically constrained are few and far between' (Rashid and Elder 2009, p. 2).

An early study of mobile phones at the base of the pyramid in Sri Lanka and India was undertaken by Zainudeen et al. (2006). They found that for financially constrained respondents with incomes of less than US$50 per month in the two countries, only 24 and 23 %, respectively, owned the mobile phones they were using. There was, however, a good deal of mobile phone sharing and use of public access points (India, in particular, has a relatively large number of these facilities). When these considerations were factored in, it turned out that the vast majority of respondents in the two countries had used a mobile phone in the three months preceding the survey.

More recently, Wendowski et al. (2012) have studied mobile phone adoption among 14 income groups in Kenya. It is unfortunate that their results are presented in graphical rather than tabular form, which in some cases make them difficult to interpret with any degree of precision. One clear and interesting result, though, is that even in the very bottom group containing the poorest individuals, there was an ownership rate of 20 %.

In the same year a study of households below the South African poverty line of roughly US$50 per month, showed that no fewer than 75 % of the population in this group have mobile phones (Infodev 2012).

The most extensive evidence of mobile phone diffusion at the bottom of the pyramid in Africa, however, has been collected through surveys by Research ICT Africa. Using a definition of the B of P as households with less than US$2.5 per household member per day, this South African based research institute has produced the outcomes for the twelve countries shown in Table 3.4.

Rates of ownership can be seen to vary from South Africa and Botswana (two relatively rich countries) on the one hand to Rwanda and Ethiopia (two of the poorest countries on the continent) on the other. At least at the extremes therefore I am led to suggest that the average income level below the poverty line is higher in richer than in poorer countries. After all, there is a substantial difference between 20 US cents and one of US$1, even though both belong to the base of the pyramid. The point can be clarified with reference to the so-called poverty gap which measures not just the number of poor people but also the average intensity of poverty below the poverty line. This gap can be technically defined as 'The mean shortfall from the poverty line (counting the non-poor as having zero shortfall), expressed as a percentage of the poverty line' (World Bank, Indicators 2013). For a given income level, the severity of the poverty gap will of course depend on the way income is distributed.

Table 3.4 Ownership of mobile phones at the BoP, selected African countries, 2011

Country	% ownership of mobile phones
Uganda	0.47
Kenya	0.65
Tanzania	0.32
Rwanda	0.22
Ethiopia	0.18
Ghana	0.51
Cameroon	0.36
Nigeria	0.61
Namibia	0.45
South Africa	0.77
Mozambique	0.42
Botswana	0.64

Source Research ICT Africa, survey data

Table 3.5 The poverty gap: selected countries, 2013

Country	Poverty gap
Ethiopia (2011)	10.4
Namibia	5.7
Nigeria	27.5
Rwanda	26.5
South Africa	1.2
Tanzania	13
Uganda	12
Cameroon	n/a
Kenya	n/a
Botswana	n/a
Ghana	n/a
Mozambique	n/a

Source World Bank Indicators (2013)

Table 3.5 displays what information is available on this gap for the twelve African countries in question. It is not surprising that the richest country, South Africa, which has the highest percentage ownership among those at the B of P, also has the smallest poverty gap (1.2 %). The main reason is that the average income of those consumers is relatively close to the line itself, thereby making a mobile phone more affordable. By the same reasoning, it is also not surprising that Rwanda, one of the poorest countries in the sample, suffers from a very much higher poverty gap. Some of the other entries, however, are more difficult to explain. Nigeria, for example, is one of the richer countries in the region, yet it exhibits the highest poverty gap in the sample countries for which data are available. Equally, as one of the poorest countries in Sub-Saharan Africa, Ethiopia might be expected to suffer from a relatively high shortfall of average income from the poverty line. Yet, at a figure of 10.4, it comes in with the third lowest gap in the table. Clearly, there are other factors than income at work in these two aberrant cases (they are aberrant, I should emphasize, from the viewpoint that country incomes determine the extent of the poverty gap).

3.2.1 Asian Evidence[2]

Note, before I move on to consider dimensions of affordability other than income, that there are several studies of mobile use at the B of P, other than those on Africa (for a 2009 review, for example, see Rashid and Elder). Perhaps the most interesting of these is the study of six Asian developing countries by Agüero et al. (2011). The authors divide incomes at the B of P into five quintiles, with the first representing the lowest income level. Table 3.6 reports on the percentage of expenditure on mobile phones for Bangladesh, Pakistan, Sri Lanka, India, the Philippines and Thailand.

In each of the six cases the percentage expenditure declines as incomes rise through the five quintiles. This trend is especially apparent in the progression from the first (poorest) quintile to the second. In India for example the change is from 24.3 to 11.3 %, resulting in a figure of 4.4 for the highest quintile. From these data Agüero et al. (2011) estimate Engel curves[3] and conclude that 'mobile phone service, for the BoP sectors of our group of countries exhibits the characteristics of a *necessary* service in economic terms, as we find income elasticities in a range of 0.1782 for the Philippines to 0.264 for India. This means that the higher the income, the lower the relative importance of mobile telephony in the individual's budget, in other words, mobile telephony expenditure is not very sensitive to changes in disposable income (mobile expenditure is inelastic with respect to income)' (Agüero et al. 2011, emphasis added).

Note that these data tell us something different about the engagement of the poor with mobile phones than the information contained in Table 3.5 for the twelve African countries. In particular, the Asian evidence tells us the expenditure on mobile phones as a percentage of B of P income, as opposed to the percentage among those at the BoP income who own such technology.

Note too that the welfare effect of spending on mobile phones at the B of P is not always unambiguously positive. For even low-cost handsets and prices may still mean that there is a high proportionate expenditure on mobile phones at the B of P, especially among the poorest groups in this category. And such expenditure,

---

[2]There is also some Latin American evidence. Galperin and Mariscal (2007), for example, find that 'the poor represent a significant market for mobile operators, with higher than expected average per capita expenditure' (p. 8). See also Barrantes and Galperin (2008) for some graphical analysis of the affordability of mobile phones.

[3]An Engel curve describes how expenditure on a particular good changes in response to a change in income.

Table 3.6 Percentage of expenditure on mobile telephone services in selected Asian countries by quintile at the B of P

Quintile	Bangladesh	Pakistan	India	Sri Lanka	Philippines[a]	Thailand
1	29.7	45.8	24.3	27.0	57.0	24.4
2	11.5	17.2	11.3	11.7	28.8	11.4
3	7.8	9.9	8.4	6.5	18.4	7.3
4	6.5	6.8	5.7	4.7	11.7	5.2
5	3.8	5.1	4.4	3.1	6.3	3.7

[a]Figures for the Philippines reflect the fact that there are respondents who do not have any income source and were assigned imputed levels
Source Agüero et al. (2011)

in turn, may imply that other necessary items in the household budget are displaced. Any such displacement, moreover, creates an imbalance in the budget, which offsets the positive welfare effect of mobile telephony on the poor.

In some cases, such as the African Millennium Project, the loss is felt in a reduction in the amount of spending on school fees, (Murphy and Carmody 2015), raising important issues of inter-generational well-being. In another case, that of Uganda, low-income groups were found to be reducing their purchase of groceries in order to buy airtime of mobile phones (Murphy and Carmody 2015). Such welfare ambiguities as these however are not the first to have been observed in the development literature. As early as 1977, for example, Wells reported on the predilection of the poor in Brazil for spending on expensive durable goods, which was causing a decline in their nutritional status (Wells 1977).[4]

3.3 Affordability

Among those who study mobile phones and development, there is quite a widespread view that affordability is often a decisive factor in the adoption decision. According to Rashid and Elder (2009, p. 3), for example, 'Affordability is a key barrier for adoption of basic mobile services by the rural poor. Approximately 50 % of the respondents in the African survey believe that the major obstacle to increased mobile use is the cost of calls'. In a similar vein the ITU (2014) argues that accessibility and use of ICTs are largely determined by the affordability of these products. Finally, Barrantes and Galperin (2008) find that affordability is the most important barrier to extending the use of mobile and their value-added

---

[4]In a perfect, neo-classical world, one could argue that the poor are revealing a preference for mobile phones over other essential goods, but in a world of advertising and imperfect information, the welfare effect is not so obvious.

services in some Latin American countries (Gamboa 2009, p. 214). A crucial aspect of affordability is price, a subject to which I turn next.

3.3.1 The Price of Mobile Services

Sub-Saharan Africa does not fare best when it comes to the price of prepaid mobile subscriptions. On the contrary, of the 20 countries with the least affordable prices in 2013, 16 were drawn from this region (ITU 2014). At the other extreme, 'Most countries with the cheapest prepaid mobile-cellular prices are in the Asia and the Pacific region, with Sri Lanka ….. and Bangladesh …. standing out with the lowest prepaid mobile-cellular prices in the world. These are examples of the levels of efficiency that operators can achieve, and how cheaper prices may help reach low-income users' (ITU 2014, p. 110).

In the first column of Table 3.7 there are prepaid price data for the twelve African countries listed in Table 3.5 and the six Asian countries described in Table 3.6. Within the former group, it is clear that there are major differences between countries. Ethiopia for example has a mobile price of just US$3.46 compared to the South-African figure of US$20.4. As argued below, such differences are due, inter alia, to variations in government tax policy towards mobile phones. But one also needs, of course, to take into account the wider regulatory framework as a whole [From this point of view, Kenya and Ghana are sometimes regarded as models for other African countries (ITU 2011)].

The second column of Table 3.7 represents the GNI per capita of countries while the final column shows the price as a percentage of the GNI per capita. These data are necessary to capture the affordability of the prices thus reported. For even an ultra-low cost represents a high percentage of income in very poor countries (Rwanda is a case in point).

Consider in this regard, the three countries in the African sample with the lowest mobile prices, namely, Ethiopia, Kenya and Ghana. These prices, in themselves, promote the affordability of the new technology for a given per capita income level. They help to explain why, as already noted, Kenya and Ghana are sometimes regarded as models of sound policy-making for countries in the region. But because these two countries enjoy only modest per capita incomes they are surpassed by some of the richer countries in the last column of Table 3.7 (Policy towards prices, that is to say, can only go so far in engendering affordability in countries with low income levels). Among those richer countries are South Africa and Botswana. Nigeria, on the other hand, does not exhibit an exceptional price level or an especially high per capita income, but its above average scores on these dimensions result in a figure that is less than 5 % in the last column i.e. one of the lowest magnitudes in the sample.

The last column of Table 3.7 shows the affordability of mobile phones with respect to average income in the sample countries. With a percentage of 0.36 Sri Lanka deserves mention in this respect because it is by far the lowest of all

Table 3.7 Mobile cellular pre-paid sub-basket, selected African and Asian countries, 2013

	Price (US$) monthly	GNI per capita (US$)	Mobile sub-basket as % GNI p.c.
African countries			
Ethiopia	3.46	470	8.83
Namibia	10.36	5.840	2.13
Nigeria	6.51	2.760	2.83
Rwanda	6.4	620	12.39
South Africa	20.4	7.190	3.4
Tanzania	9.37	630	17.85
Cameroon	17.67	1.270	16.7
Kenya	3.84	930	4.96
Botswana	10.57	7.730	1.64
Ghana	5.15	1.760	3.51
Mozambique	12.74	590	25.91
Uganda	9.41	510	22.14
Average	9.7	2.525	10.2
Asian countries			
Bangladesh	1.41	900	1.88
Pakistan	3.65	1.380	3.17
India	2.91	1.570	2.23
Sri Lanka	0.95	3.170	0.36
The Philippines	10.15	3.270	3.72
Thailand	5.36	5.370	1.2
Average	4.1	2.610	2.1

Source ITU (2014)

the estimates in the table and despite an income that is far from being exceptionally high, this country is ranked twelfth in the world (ITU 2014). In terms
of the last column of Table 3.6, this evidence helps to explain why Sri Lanka
has the lowest number in the Asian sample in the majority of the B of P quintiles shown there. More generally, it is clear from Table 3.7, that affordability
among the Asian sample countries is around five times greater than in Sub-
Saharan Africa (2.1 as against 10.2 %).

3.3.2 The Cost of Mobile Handsets

Another factor that impinges on the degree to which the poor can afford to own
mobile phones is the price of the handset. This is partly a matter of the growing availability in developing countries of low-cost handsets that are often

Table 3.8 Taxes as a % of handset cost (selected regions)

Region	% descending order
Sub-Saharan Africa	29
Latin America	27
Central and Eastern Europe	25
Middle East and Maghreb	21
EU	20
Asia Pacific	17

Source GSM (2011)

manufactured in China. These products are available in some such countries at around US$10 to US$15 (Chick et al. 2010).

But the cost of mobile phones to the poor also depends on the extent of taxes imposed on the purchase of these goods by governments in developing countries. Table 3.8 shows taxes as a proportion of handset costs for some of the major regions in the world (where such taxes include customs duties on imported products and a variety of other handset-specific charges such as customs duties and luxury taxes).

It is unfortunate that the tax percentage is highest (29 %) in the region where on average penetration is lowest. It is even more undesirable that some of the African countries in which mobile phones have spread the least at the B of P, are afflicted with taxes that are higher than the African average. I am thinking here for example, of Cameroon (48.25 %) and Rwanda (48 %). At the other extreme, however, lies Kenya, with one of the lowest tax rates (2.25 %) on mobile handsets in the world. This is partly the result of the abolition of VAT on mobile phones in 2009 (Chick et al. 2009) in that country. Kenya's relative success with the spread of these products at the B of P, also has to do, I should emphasize, with the prevalence there of the M-Pesa mobile banking application (see Chap. 5 below).

For the Asian six country sample described above, taxes as a percentage of mobile handsets are set out in Table 3.9.

The countries in Table 3.9 divide quite readily into two groups: one headed by Sri Lanka (with no tax on handsets) and includes Thailand and India with relatively low taxes. The other may be described as the high-tax group consisting of

Table 3.9 Taxes as a % of handset costs (selected Asian countries), 2011

Region	%
Bangladesh	33.75
Pakistan	36.08
India	13.56
Sri Lanka	0.0
Philippines	22.0
Thailand	12.0

Source GSM (2011)

Bangladesh, the Philippines and Pakistan. These inter-country patterns are clearly mirrored in and indeed partly underlie the B of P results shown above in Table 3.6. It seems, moreover, that in general the Asian countries with low mobile prepaid prices also tend to be among those with low taxes on handsets and vice versa, leading to a cumulative type pattern of country behaviour at the B of P (though I should point out that Bangladesh is somewhat exceptional in this regard, exhibiting a very low mobile price (see Table 3.7) but a comparatively high rate of tax on handsets). So far I have confined myself to the degree to which the poor engage themselves with mobile phones via adoption of this technology. But the impact on this group also obviously has implications for inequality. I therefore complete this chapter with a brief discussion on the impact of mobile penetration on inequality in developing countries (assuming for the moment that such technology has a positive effect on the income of its users).[5]

3.4 The Impact on Inequality

Recall from the previous discussion of the Rogers model of technology diffusion, that the adoption of an innovation takes place initially among the more affluent members of society within and between countries. In the beginning therefore new technologies in general and mobile phones in particular, are a force making for inequality.

After a time, however, the new product or technology falls in price and tends, partly as a result, to be spread more widely among less affluent members of the society and becomes instead a force making for greater equality, though one should note here that most innovations in developing countries do not progress beyond the 25 % level of penetration (World Bank 2008). Mobile phones are an exception in this regard because, as this chapter has shown, they have spread widely at the B of P in numerous poor countries. As a result, they ought to have effected quite a large reduction in inequality in those (and certain other) countries. Asongu (2012) has empirically confirmed this expectation, by showing, for a sample of 52 African countries, that the mobile penetration rate is significant in a regression equation with the Gini coefficient as the dependent variable. 'The findings suggest that mobile penetration is good for the poor, as it has a positive income-redistributive effect' (Asongu 2012, p. 14). As the author sees it there are a number of aspects to this.

> Firstly, many lives have been transformed by the mobile revolution thanks to basic financial access in the form of phone-based money transfer and storage…. Hence, the significant growth rates of mobile telephony that is transforming cell phones into pocket-banks in Africa is providing countries in the continent with increased affordable and cost-effective means of bringing on board a large part of the population that have until now been excluded from formal financial services for decades.

---

[5]See Chap. 6 below for a discussion of unique mobile subscribers for a range of Asian Countries.

Secondly, mobile phones can assist households' budget when faced with unpredictable shocks which drive poverty. The probability of a poor family incurring drastic loss due to an unpredictable shock is certainly mitigated and lowered when families are able to respond .... In a more timely fashion.

Thirdly, mobile phones could empower women to engage in small businesses (and/or run existing businesses more efficiently), hence enabling them to bridge the gap between gender income inequality (Asongu 2013, p. 15).

Note, though, that there are still a number of countries in Africa, (and elsewhere), where mobile phones have spread to the B o P only to a very limited degree (see, for example, the Ethiopian case above in Table 3.5). In these cases, the propoor effects of the technology will have done little to counter the more widespread adoption at the upper echelons of the pyramid as a whole.

3.5 Conclusions

As its point of departure, this chapter has argued that in order to understand the impact of mobile phones on poor people in poor countries, it is first necessary to establish how many people in this group are actually affected by these products. I began investigating this prior question on the basis of international statistical data on prepaid subscriptions as a percentage of the population. This, however, turned out to be a rather unprofitable line of enquiry because the data in question overestimate the involvement of the population in adopting mobile phones. For what the conventional data about subscriptions involve is more about SIM cards per head than what are known as unique mobile subscribers. Thus, even when 100 % of the population enjoys a subscription, not everyone is connected to a mobile network (indeed, it may be only 20 % of all individuals).[6]

I turned next to more direct estimates of the involvement of the poor with mobile phones. Two pieces of survey evidence were of particular importance in this regard. One of them is about adoption at the B o P in twelve African countries and the other concerns behaviour among this group in six Asian nations. Apart from some notable exceptions, both samples revealed a quite considerable involvement of the poor in ownership and use of mobile telephony. Country differences were ascribed, tentatively, to deviations in income levels and government policies towards prices and taxes on handsets.[7]

---

[6]GSMA (2014) contains estimates of unique mobile subscribers for a range of Asian Countries.

[7]It is often not adequate to focus only on handset taxes. The World Bank (2012) for example points out the negative effect of taxes on SIM cards in Bangladesh, which, as noted above has one of the lowest mobile prices in the world. 'The tax of TK 800 ($11,60) on new SIM cards has a huge negative impact on low-end subscribers' (World Bank 2012, p. 119).

References

Agüero A, de Silva H, Kang J (2011) Bottom of the pyramid expenditure patterns on mobile services in selected emerging Asian countries. Available via http://itidjournal.org/itid/article/viewFile/758. Processed 10 April 2013

Aker J, Mbiti I (2010) Mobile phones and economic development in Africa. J Econ Perspect 24(3):207–232

Asongu S (2012) The impact of mobile phone penetration on African inequality. Available via https://mpra.ub.uni-muenchen.de/46041/. Accessed 12 July 2013

Barrantes R, Galperin H (2008) Can the poor afford mobile telephony? Evidence from Latin America. Tel Pol 32:521–530

Chick C, Nique M, Smith F, Taverner D (2010) Increasing rural mobile connectivity. GSMA development fund. Available at http://gsmag/globaltaxreviewnovember2011.pdf. Accessed 3 Feb 2015

Galperin H, Mariscal J (2007) Mobile opportunities: poverty and telephony access in Latin America and the Caribbean. Available at https://dirsi.net/files/regional_FINAL.pdf. DIRSI/IDRC Accessed 5 Jan 2015

Gamboa L (2009) Strategic uses of mobile phones in the B of P: some examples in Latin American countries. Lect de Economia 71:209–234

Gillet J (2014) Measuring mobile penetration. GSMA intelligence. Available via https://gsmaintelligence.com. Accessed 10 Dec. 2014

GSMA (2011) Global mobile tax review 2011. Deloitte Available at http://gsmaglobaltaxreviewnovember2011.pdf. Accessed 3 Feb 2015

GSMA (2014) The mobile economy 2014. Available at http://www.gsmamobileeconomy.com. Accessed 9 Jan 2015

InfoDev (2012) Mobile phone usage at the base of the pyramid in South Africa. Available via: https://www.infodev.org/infodev-files/final_south_africa_bop_study_web.pdf. Accessed 27 October 2013

ITU (2011) Measuring the information society, Geneva

ITU (2014) Measuring the information society report. Geneva

Murphy J, Carmody P (2015) Africa's information revolution. Wiley-Blackwell, Hoboken

Rashid A, Elder L (2009) Mobile phones and development: an analysis of IDRC-supported projects. EJISDC 36(2):1–16

Wells J (1977) The diffusion of durables in Brazil and its implications for recent controversies concerning Brazilian development. Cam J Econ 1:259–279

Wesolowski A, Eagle N, Noor A, Snow R, Buckee C (2012) Heterogeneous mobile phone ownership and usage patterns in Kenya. doi: 10.1371/journal.pone.0035319

Wheeler D, Dasgupta S, Lall S (2001) Policy reform, economic growth, and the digital divide: an econometric analysis. World Bank policy research working paper, Washington D.C. Available via http://dx.doi.org/10.1596/1813-9450-2567. Accessed 12 Oct 2003

World Bank (2008) Global economic prospects. Available via: http://econ.worldbank.org/WBSITE/EXTERNAL. Accessed 25 Jul 2011

World Bank (2012) Maximizing mobile. Washington D.C

World Bank (2014) The little data book on information and communication technology. Washington D.C

Zainudeen A, Samarajiva R, Abeysurival (2006) Strategic use of telecom services on a shoestring: strategic use of telecom services by the financially constrained in South Asia Available via LIRNEasia. http://www.lirnesia.net/2006/02/strategic-use-of-telecom-services-on-a-shoestring. Accessed 17 April 2010

Chapter 4
A Pro-poor Bias: Leapfrogging and the Context

Abstract The previous chapter suggested that the mobile phone has spread quite widely at the B of P even in relatively poor developing countries. I argued there that the popularity of this technology had to do partly with its affordability. In this chapter I seek to further explain the popularity of the mobile among those living at the BoP and to discuss how they benefit from it. My contention is that these issues have partly to do with the nature of the technology itself and partly with the context in which it is introduced in relatively backward areas of developing countries. In both cases, I contend that there is evidence of a pro-poor bias in the way benefits accrue to the various income groups.

Keywords Appropriate technology · Beeping · Sharing · Leapfrogging · Network effects

I turn first to a discussion of the technology itself.

4.1 The Technology

The discussion on technology is divided into two parts. The first deals with the leapfrogging characteristics of mobile phones (where leapfrogging is broadly defined as the process of bypassing an earlier technology and moving directly to the new one). The second aspect has to do with the characteristics of mobile telephony that make it appropriate to the poor.[1]

---

[1]For a general discussion of appropriate technology see Stewart (1977).

© The Author(s) 2016
J. James, *The Impact of Mobile Phones on Poverty and Inequality in Developing Countries*, SpringerBriefs in Economics,
DOI 10.1007/978-3-319-27368-6_4

Table 4.1 Fixed telephone
subscriptions 2013 (selected
African countries)

Country	Subscribers per 100 people
Botswana	9
Kenya	0
Tanzania	0
South Africa	9
Ethiopia	1
Uganda	1
Chad	0
Namibia	8

Source World Bank, world development indicators, power and
communication, 2015

4.1.1 Leapfrogging Characteristics of Mobile Phones

According to the Economist (2008), 'The mobile phone is also a wonderful example of a "leapfrog" technology: it has enabled developing countries to skip the
fixed-line technology of the 20th century and move straight to the mobile technology of the 21st.' Indeed, according to this same publication,

> The mobile turns out to be rather unusual. Its very nature makes it an especially good leap-
> frogger: it works using radio, so there is no need to rely on physical infrastructure such as
> roads and phone wires; base-stations can be powered using their own generators in places
> where there is no electrical grid; and you do not have to be literate to use a phone, which is
> handy if your country's education system is in a mess (The Economist 2008).

This quotation does not however draw attention to the encumbering effect of the
older technology on the diffusion of the new: in this case the effect of the existence of
fixed phones on the adoption of mobile. The point is that 'some industrializing economies are less hampered by commitments to previous generations of technology.
Developing countries … may benefit from the windows of opportunity provided by
the new paradigm, especially at the early stages of diffusion' (Hobday 1995, p. 137).[2]

Thus stated, this thesis has a close affinity to the theory of catch-up, advanced
in 1962 by Gerschenkron, namely, that developing countries with the least commitment to the earlier forms of technology have the most to gain from diffusion of
the new (Gerschenkron 1962). The reason for this is the fact that in such countries
mobiles are chiefly serving as substitutes for fixed lines (as opposed to the com-
plementary role they play in developed countries) and their value in fact derives
substantially from the absence of the latter.

Consider, from this point of view, Tables 4.1 and 4.2 which show, respectively,
the adoption of fixed-line phones in a selected sample of African countries and the
ratio of mobile to fixed phones in the world's major income groups. The former
table indicates that the earlier technology is almost only adopted by the richer
countries, South Africa, Botswana and Namibia. The remaining countries have

---

[2]See James (2009) on the problem of measuring leapfrogging and a suggested new measure.

Table 4.2 Fixed and mobile subscriptions by income category, 2013

Income group	Fixed phone	Mobile phone	Ratio of mobile to fixed
	(Subscriptions per 100 people)	(Subscriptions per 100 people)	
Low-income	1	53	53
Middle-income	12	92	7.7
– Lower middle	5	85	17
– Upper middle	19	100	5.3
High income	42	120	2.9

Source World Bank, World Development Indicators, Power and communications, 2015

negligible or zero rates of diffusion. Table 4.2 shows clearly that the ratio of mobile to fixed phones varies inversely with income levels. And if that ratio can be regarded as a simple measure of leapfrogging, the poorest countries have been by far the most successful (further disaggregation along regional lines would indicate that the majority of those countries are located in Sub-Saharan Africa).[3]

This result implies that the gains from leapfrogging will be most intensely felt by low-income countries and by extension of the argument to low-income persons within those countries as well. If the extent of transport infrastructure is also inversely related to income levels, as well it might be, the pro-poor bias of mobile phones will be accentuated. I shall return to this theme in the following chapter.

4.1.2 Mobile Phones as an Appropriate Technology

The notion of appropriate technology, or more specifically, appropriate products, is based on the idea that goods should be viewed as 'bundles' of characteristics (such as the taste and convenience of different foods). Preferences for different characteristics are assumed to depend on incomes of the household or individual. As these rise, preferences are thought to move away from basic or functional characteristics towards more 'luxurious' or non-functional ones (such as for status).[4] Appropriate products are those which are designed to meet the demands of poor users in developing countries, as shown in Fig. 4.1.

The horizontal axis is labelled 'essential' characteristics and the vertical refers to 'luxury' or 'non-essential' attributes. Two goods, X and Y, are represented in the figure and it is assumed that if the user spends all his or her budget on Y, point B can be reached, while in the case of X the attainable point is A. Whether the consumer chooses X or Y depends on the shape of his indifference curves: if they resemble IC′ and IC″, A will be chosen, whereas in the case of the curves IC′

---

[3]See James (2009) for relevant data and discussion.

[4]This argument is spelt out in Stewart (1977). In brief, one can think of preferences for product functionality giving way to preferences for say status and style.

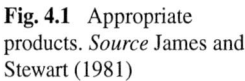

Fig. 4.1 Appropriate
products. *Source* James and
Stewart (1981)

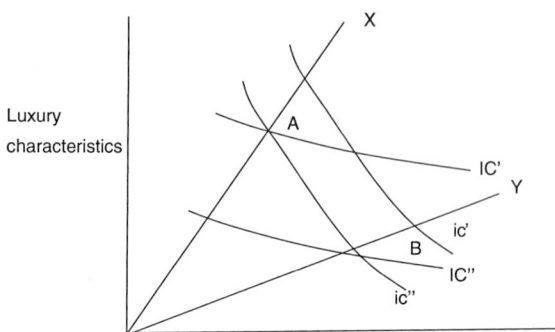

and IC″, B will be selected. Low-income users will thus tend to favour product
Y, while more affluent ones will instead favour X (in each case the selection of
A or B will put the adopter on a higher indifference curve). For obvious reasons,
the former is referred to as an appropriate technology or product while the latter is
deemed to be inappropriate to the needs of the poor.

Not all mobile phones conform to product Y in the figure. On the contrary,
many of them would be better described as X, the inappropriate version. This is
because they embody characteristics that are superfluous to the main function of
the product, which is to communicate. These additional characteristics (see immediately below) push up the cost of the mobile phone so that it becomes unafford-
able by the poor.

At the same time, though, there are many appropriate mobile phones, often
from China and India and often embodying low-cost handsets. They do not con-
tain characteristics such as cameras, Internet connectivity or substantial amounts
of storage space. But they frequently do embody a lot of battery capacity for those
without access to electricity in poor countries. According to Castells et al. (2007)
it is 'the flexibility of wireless communication technology' that has allowed 'the
development of a variety of techniques to adjust mobile technology to developing
countries' (Castells et al. 2007, p. 218).

The same authors provide perceptive insight into the appropriateness of mobile
phones in developing countries, though they do not express their argument in those
terms. Rather, in their own words:

> In developed countries, mobile phones are defined by the term *mobile* and appreciated as
> a means to communicate on-the-go. However, the immediate benefit for people in devel-
> oping contexts is that of *connectivity*, associated with having a means of communication
> whether mobile or not. Thus, considerations linked to mobility, style, and Internet access, for
> example, are arguably secondary to that of basic connectivity at this stage …. It is therefore
> important to identify the true source of developmental benefit. For people who already own a
> landline at home and/or at work, mobile phones bring the added benefit of mobility and con-
> venience. To those for whom the mobile phone is the first form of personal communication
> to be owned, the major prize is to be connected at last; the phone is acquired not in order to
> be mobile, but in order to be connected, although mobility is an added bonus. This is prob-
> ably the major difference between mobile-phone owners in areas with fixed-line teledensity
> and those with low levels of fixed lines in developing countries (Castells et al. 2007, p. 218).

For these reasons, it is not uncommon to find mobile phones being used effectively as fixed lines in developing countries, with a permanent location in the household and sometimes a place where it is permanently plugged in Castells et al. (2007).

4.2 The Context

Three key points need to be emphasized at the outset of the following discussion on the specific components of the context in which mobile phones are used by the poor in developing countries.

The first is that the context does not somehow emerge in an exogenous form in these (or developed) countries. Rather, it is the product of active and continuing involvement of many actors in the specific socio-cultural-economic way of life (the case of mobile banking discussed below is perhaps the clearest example of this point). The second, and related recognition, is that the context which governs mobile phone use among the poor in developing countries is very different from that which confronts users in the developed world. In too many cases, however, this distinction is ignored by authors who, and institutions which, extrapolate the latter context to the former, thereby ignoring all the specifically developmental components of the issue (see below, for example, the case of sharing mobile phones which is prevalent in many parts of the Third World, but which scarcely exists in the First World, where, by contrast, most everyone owns at least one mobile phone).

The final recognition is that the view of a context offered here bears some resemblance to Sen's concept of functionings (Sen 1985). That concept posits that what matter for welfare are not goods themselves, but rather, what people are able to accomplish with them. For example, in the case of food, the conversion of nutrients into nutritional well-being, depending among other things, on access to medical services and climatic conditions (Sen 1985). Thus, even an apparently narrow physiological process comes to depend on the broader environment.

I seek now to show that there are important components of the context in developing countries which exert a pro-poor bias in the distribution of gains from use of the mobile phone.

4.2.1 Network Effects

These are basically benefits that accrue to the members of a network from the addition of extra participants. Each person, that is, is favourably affected by the behaviour of others. Because these benefits usually go unpaid, they are described in the economics literature as "positive externalities" (Easley and Kleinberg 2010). Network effects are commonly experienced in social network groups of various kinds where there is interaction among the participants. A natural setting where network effects arise is in the adoption of technologies for which interaction or compatibility with others is important. For example, 'when the fax machine was first introduced as a product, its value to a potential customer depended on how many others were also using the same technology. The value of a serial networking site exhibits the same properties: it's valuable to the extent that other people are using it as well' (Easley and Kleinberg 2010, p. 509).

In the case of mobile phones the situation is similar: the benefits to an individual subscriber depend on how many others also have access to the technology. But in a more precise definition, the user of a mobile may be more interested in the adoption behavior of his or her immediate social grouping (made up primarily by family and friends).[5]

An interesting question is then whether the grouping that is relevant to the poor is larger or smaller than for other income categories. It is sometimes argued for example that poverty is associated with social isolation (Douglas and Isherwood 1979), I know of no empirical evidence however that addresses this question in the context of developing countries.

From another point of view, however, Castells et al. (2007) argue convincingly that network effects tend to be larger for poor than rich countries. Their case is similar to the one laid out above for the benefits of leapfrogging for the two groups of countries. As they see it, 'Mobile phones provide significantly higher network effects in developing than in developed countries where fixed lines have already performed this function' (Castells et al. 2007, p. 216). The authors go on to cite Waverman et al. (2005, p. 17) to the effect that 'while in developing countries the benefits of mobile are two-fold-the increase in the network effect of telecoms *plus* the advantage of mobility-in developed economies the first effect is much more muted'.

4.2.2 Sharing Mobile Phones

In developed counties, as already noted, mobile phone ownership is the principal means by which benefits therefrom are extracted. In developing counties on the other hand, mobile technology will often tend to be beyond the means of the poorer members of the BoP. One major institutional response to this distinction between rich and poor countries has been the practice of mobile sharing in the latter. Indeed, in many regions of Africa and Asia, sharing has become quite a widespread phenomenon (Aker and Mbiti 2010; James 2013). One study of Rwanda, for example, found that fully one third of survey respondents shared their phone

---

[5]According to Birke and Swann (2006, p. 3) 'especially in markets with direct interaction between consumers, like mobile, telecommunications, it is rather an individual's social network that determines an adoption decision'.

with family and friends. Some see in this behaviour a reflection of a 'culture of sharing' in developing countries, referring, for example, to the widespread practice there of sharing newspapers (James 2011). Because they are a good deal more expensive than a newspaper, so the argument runs, the sharing of mobile phones should be at least as pronounced, especially among the poor.[6]

Another reason for the relatively intensive use of mobile sharing in developing countries has to do with household size. The point is that,

> Access is particularly high in countries with large households. Take Senegal, where the subscription penetration was 57 per 100 people in 2009, but household penetration was estimated to be 30 points higher at 87. This larger household size can dramatically extend access to mobile phones, considering that on average nine persons are in each Senegalese household. Several low-income nations have higher mobile phone penetration than some developed economies. For example, Senegal, along with some other low- and middle-income economies, has a higher proportion of homes with mobile phones than either Canada or the United States (Minges 2012, p. 116).

Large-scale families also tend to be relatively poor and to this extent, sharing will tend to favour this group over those that are more affluent. Note, though, that sharing may carry a different welfare impact as compared with owning. For, in the former case incoming calls will tend to be answered by the owner who has to decide whether or not to alert the family member for whom the call is intended. Plainly, if this does not occur, the phone loses part of its positive welfare impact as compared with the ownership case.[7]

Note finally that sharing has been known to play a role in gaining access to electricity for recharging mobile phones. According to Samuel et al. (2005, p. 52), for example, 'Even the absence of electricity does not present an insurmountable barrier, thanks to the sharing of mobiles and recharging batteries in the nearest town, or, recharging locally by a generator or car battery'.

4.2.3   Prepayment

It is generally agreed that pre-, as opposed to post-paid services have done much to make mobile phones accessible to the poor in developing countries. Galperin and Mariscal, for example, 'confirm the importance of prepaid service models for a population with fluctuating incomes and limited insertion in the formal economy. Overall, low-income mobile users prefer prepaid plans' (Galperin and Mariscal 2007, p. 4). These authors have, in fact, done the most to relate prepaid call services to the BoP in a developing country context.

Basing their results on a large-scale survey of low-income households in Argentina, Brazil, Colombia, Jamaica, Mexico, Peru and Trinidad and Tobago,

---

[6]Though there may arguably be more reluctance on the part of mobile phones owners to part with their goods, as compared to the owners of newspapers.

[7]The extent to which this effect would detract from the user's welfare, however, is unknown.

Table 4.3 Preference of the poor for prepaid plans, selected countries in Latin America

Country	Cost control	Prepaid is cheaper
Trinidad and Tobago	61	29
Brazil	52	38
Argentina	47	35
Mexico	46	26
Colombia	41	31
Jamaica	41	32
Peru	34	33

Source Galperin and Mariscal (2007, p. 5)

they find that the preference for prepaid plans can be ascribed to two main factors. 'The main motive is spending control: users value the ability to purchase credit when they have cash in hand, rather than committing to a fixed monthly charge. The other factor is cost: users perceive prepaid plans to be cheaper than post-paid plans' (Galperin and Mariscal 2007, p. 4).

Table 4.3 shows the relative importance of these two factors in the seven countries concerned.

For all countries in the table, therefore, it is the capacity to control mobile phone spending (pay-as-you-go), that is more important than the perception that prepaid plans are cheaper, though in Peru the difference is slight. Both reasons, however, contribute to the fact that pre-paid is the overwhelming choice of mobile plans among inhabitants of poor countries, especially those that can be described as living in poverty (World Bank 2012; for African evidence see household surveys by Research ICT Africa).[8] According to data provided by Gillwald (2005), households in Sub-Saharan Africa with a prepaid system, have an average income of US$487, as against the figure of US$1911 for those who use a postpaid plan.

4.2.4 Rental Markets

Another institution that caters to the mobile needs of the poor in developing countries is the rental market, which, unlike some family sharing arrangements, does invariably require a payment for time used. In other words, it is a strictly commercial operation as compared with family sharing which may often be free.

One of the most prominent examples in this regard is the Grameen Telecom initiative in rural Bangladesh, which is based on the idea of Village Pay Phones (VPPs) run by women who operate mobile phones from somewhere in the middle of the village. From small beginnings this initiative has spread to more than half the country's villages, affecting tens of millions of rural inhabitants. There are numerous studies of this highly successful program (e.g. Bayes et al. 1999;

---

[8]See for example RIA (2008) for some early research on the issue.

Table 4.4 Alternatives to VPPs (% of respondents)

	Economic status		
	Extremely poor	Moderately poor	Non-poor
Would not try	–	–	2.3
Telephone from other phone	26.5	36.5	43.0
Post office	5.9	7.1	6.8
Have to go/hire person to go	67.6	56.4	47.3
Other	–	–	0.6
Total	100.0	100.0	100.0

Source Bayes et al. (1999)

Richardson et al. 2000) but from my point of view one stands out in importance because it explicitly relates the Grameen experience to the poor in Bangladesh. The study in question is by Bayes et al. (1999) and it contains a discussion of the impact of VPPs on the welfare of different income groups.

It shows first of all what the alternative would be without mobile phones for each such group. In particular respondents were asked: 'How would you meet the purposes of the current calls had there been no VPP in your village?' (Bayes et al. 1999). Table 4.4 shows the responses classified by income group.

For all three groups, the most common alternative is that one would either go or hire someone to go to the location in question. But the percentage is highest among the poor (at 68 %). 'In other words, if no VPPs had been available, the poor would have required greater physical mobility than the non-poor. ... By and large it seems that the absence of VPPs would inflict relatively more transaction costs for communications on the poor' (Bayes et al. 1999, p. 28).

This result is important because it underlies that study's findings with regard to the consumer surplus that accrued to the different income groups as a result of the VPP initiative (the consumer surplus is the difference between what the user would pay and what he actually pays). The main finding in this regard is that the consumer surplus of the poor (a measure of welfare) is 50 % higher than that of the non-poor (As an indicator of the amount that the consumer surplus of the former group represents, the authors note that it would buy 12 kg of coarse rice). 'That the poor would reap the maximum CS following the advent of VPPs according to the Bayes study, is 'quite obvious'. For, 'the poor usually do not have much in the way of alternatives to communicate with the outside world, neither relatives to help with a phone call, nor relatives to provide a ride to the destination. For the poor, the advent of VPPs opened up a lower-cost alternative for exchanging information' (Bayes et al. 1999, p. 29).

The reader will notice here a distinct similarity with the role played by fixed phones in determining the impact of mobile phones on the poor. In particular, just as the unavailability of those phones tends to enhance the welfare impact of mobile telephony, so too does the lack of other communication options benefit the poor relative to higher income groups. Here as well, therefore, there is a discernible pro-poor bias in the gains from this form of telephony among different income groups.

Another successful example of a mobile rental business is Vodacom's Community Services Programme in South Africa. It is successful in the sense that the original idea has rapidly been scaled up to the national level (for a discussion of the transition from the micro to the macro level see Chap. 6). The programme is relevant also because it shows 'how a technology company can learn to operate profitably in a lower-income market segment' (Porritt 2005, p. 253).

Vodacom operates as a joint venture between Vodafone and Telkom, South Africa. It has generated a 'shared service model' for supplying mobile phone (and telecommunication services more generally) to low-income communities. 'The basic approach has been to set up stationary phone shops or kiosks with multiple lines, all connected to Vodacom's existing infrastructure through a wireless link' (Porritt 2005, p. 253). Some of the kiosks take the interesting form of used shipping containers, which, presumably, serve as a low-cost, appropriate technology in the sense defined above.

More generally, operator-owned or leased phone shops with mobiles to rent, are a common feature of many street corners in Africa and elsewhere (In Botswana, for example, mobile phone kiosks are responsible for almost 85 % of public phone calls).[9] They are in effect fixed mobile payphones that do not allow return calls to be made to customers. As such, they tend to impart a less favourable welfare impact on the poor than does the Grameen model (in which customers are alerted when they receive incoming calls). Much though depends on the specific context, as Samuel et al. (2005) have pointed out. More precisely, they draw a distinction between two types of communication. To wit,

> Communications which are initiated by an individual to acquire data or information from a central source (such as finding out the availability of goods in a shop) are largely unaffected by the inability for the individual initiating the call to be reached in turn (Samuel et al. 2005, p. 40).

On the other hand, there are communications where the ability to receive return calls can be crucially important. Samuel et al. (2005) refer in this context to calls that were made in order to find a job. These entailed not only attempts to gain information and making an application but also served as a means of being contacted by potential employers: 'that is, inbound communication was important', (Samuel et al. 2005, p. 40).

4.2.5 Beeping

Mobile phone users can sometimes engage in an interesting practice of beeping or intentional missed calls. I will not attempt here to describe all the intricacies

---

[9]See James (2013) who also provides a discussion of South Africa's Community Phone Shops.

of this often subtle form of interaction, but will focus instead on those aspects that bear most directly on relatively poor users of mobile telephony [on the more anthropological aspects of the phenomenon, see Donner (2007)].

The first type of beeping in this context occurs when an unfinished call conveys a signal to the recipient. The message may take the form, for example, of an agreed-on upon meaning such as: two rings mean 'meet me at a certain time' or 'I am leaving in ten minutes' (Donner 2007). The advantage to the notional poor user who initiates the call is that it is practically costless (and does not require the ability to receive incoming calls).

The second form of beeping differs from the first in that there *is* an expectation of a return call, but it remains free to the relatively poor individual who initiates the beep. For, in some settings observed by Donner (2007), such as Rwanda, there exists a communal consensus in which the relatively rich are expected to return the missed call. 'There are clear conventions as to who should and should not beep' (Donner 2007). These can often be summarized by the expression 'the richer guy pays'. To this extent, beeping is brought to bear as an instrument that favours relatively poor users of mobile phones (but only those with an incoming call facility). This tendency is offset to some extent, however, by the fact that certain forms of mobile use, such as rentals or sharing, do not permit the reception of incoming calls as discussed above. But for ownership modes of access this is obviously not a problem. The more do the poor rather than the rich engage in rental and sharing arrangements relative to ownership, the more severe will tend to be this countervailing effect.

The evidence on beeping by income group is unfortunately very scarce. One of the few available pieces of data concerns Rwanda and in particular that in that country respondents with lower income are more likely to make missed calls compared with more affluent users (Rashid and Elder 2009).

4.2.6 Mobile Banking

The World Bank (2012) clearly describes the typical financial context for the poor before the introduction of mobile banking in developing countries. That situation was typically one in which access to the formal banking system was exceptionally difficult for a number of reasons, including a lack of collateral.[10] As a result, the poor were unable to undertake measures to improve productivity, to start small-scale firms and farms, or to invest in health and education. Transferring funds required a visit to a physical location, which meant having to rely on a scarce and typically unreliable transport infrastructure.

The most conspicuous and successful response to this undesirable financial context was Kenya's now famous mobile banking system called M-PESA, which

---

[10]For a useful and detailed discussion of financial markets in developing countries see Ray (1998).

was initiated in that country in 2007 by 'Safaricom', a subsidiary of the British multinational Vodafone.

M-PESA is an electronic payment and store of value system that is accessible through mobile phones. To access the service, customers must first register at an authorized M-PESA retail outlet. They are then assigned an individual electronic money account that is linked to their phone number and accessible through a SIM card-resident application on the mobile phone. Customers can deposit and withdraw cash to/from their accounts by exchanging cash for electronic value at a network of retail stores (often referred to as agents). Once customers have money on their accounts, they can use their phones to transfer funds to other M-PESA users and even to non-registered users, pay bills and purchase mobile airtime credit (Mas and Radcliffe 2010, p. 1).

I will return to discuss the reasons for this successful[11] venture in Chap. 6, but the main point here is that its huge popularity in Kenya ought, surely, to have acted as a spur to the further adoption of mobile phones. After all, M-PESA does nothing less than revolutionize the financial aspect of people's lives in that country and elsewhere in the developing world.

There is certainly some hard evidence that the lack of a mobile phone is by far the most important reason for not using M-PESA (Jack and Suri 2010). Since the project's inception, though, this constraint has become progressively less binding as mobile phone adoption spread rapidly through the population. By 2012, 93 % of Kenyans could be counted as mobile phone users and 73 % as mobile money clients (Demombynes and Thegeya 2012).

Apparently there is multiple causality: the existence of mobile banking enhances the appeal of mobile telephony, which, in turn, permits that form of banking to occur. Arguably it is the poor who have gained the most from mobile banking since they were the most disadvantaged by the former system described above, but there is, as yet, no evidence to support such a claim.

4.3 Conclusions

In the previous chapter I suggested that the mobile phone has been quite widely diffused at the BoP even in some relatively poor developing countries. The purpose of this chapter was to further explain the popularity of mobile telephony among those at the BoP and to discuss how they benefit from it. The discussion was divided into two parts: one dealing with the technological aspects of the mobile phone that make it suitable for people with low incomes and the other describing how the nature of the environment (social-cultural-economic) influences the adoption of this technology by such persons and the extent of benefits that accrue to them. The results are summarized in Table 4.5.

---

[11]It is indeed one of the most successful projects in the entire ICT experience in developing countries.

Table 4.5 Summary of chapter

Subject	Mechanism	Benefits to poor
Technology Leapfrogging characteristics	Gives poor countries opportunity to leapfrog fixed-line phones	Easier adoption of mobiles, higher benefits from adoption by poor; pro-poor bias
Appropriate technology (AT) characteristics	Technology embodies characteristics for poor	Suitable for adoption by poor eq, low-cost handsets; focus on connectivity; pro-poor bias
The Context Network effects	Value to individual depends on adoption by others	Encourages adoption of mobile; effects greater in poor rather than rich countries
Sharing mobile phones	Use of phones by friends/family	Gives (sometimes free) access to the poor who cannot afford ownership; pro-poor-bias
Prepayment	'pay-as-you-go' schemes	Favours adoption of mobile phones among poor by controlling and reducing telephone costs; relates to AT
Rental markets	Buying airtime instead of buying phone	Gives access to poor unable to own mobile phones; pro-poor bias in gains from use
Beeping	Deliberately making unfinished calls	Allows poor to make costless calls; pro-poor bias
Mobile banking	Transferring money and paying bills	Encourages adoption of mobile; allows access to formal banking

The conclusion is that with respect to both the technology and the context in which it operates, there are forces that tend to favour[12] those at the BoP. Some of those forces inhere in the technology but many of them result rather from the deliberate and interactive efforts of actors to alter the context. I am referring here for example to prepaid schemes, mobile banking and beeping. In any case though, that context is one which is markedly different from the one that prevails in developed countries.

References

Aker J, Mbiti I (2010) Mobile phones and economic development in Africa. J Econ Perspect 24(3):207–232

Bayes A, Von Braun J, Akhter R (1999) Village pay phones and poverty reduction. Discussion paper on development policy, ZEF Bonn

---

[12]It should be realized, though, that not all contextual factors favour the poor. For example, with shortages of electricity it is difficult to recharge mobile phone batteries. In a later chapter I discuss some possible solutions to this problem. For some interesting examples in Tanzania and South Africa, see Samuel et al. (2005).

Birke D, Swann G (2006) Network effects and the choice of mobile phone operator. J Evol Econ 16(1):65–84

Castells M, Fernandez-Ardevol M, Qui J, Sey A (2007) Mobile communication and society: a global perspective. MIT Press, Cambridge MA

Demombynes G, Thegeya A (2012) Kenya's mobile revolution and the promise of mobile savings. World Bank policy research working paper no. 5988, Washington DC

Donner J (2007) The rules of beeping: exchanging messages via intentional 'missed calls' on mobile phones. J Comp Med Comm 13(1):1–22

Douglas M, Isherwood B (1979) The world of goods. Basic Books, New York

Easley D, Kleinberg J (2010) Networks, crowds and markets. Cambridge University Press, Cambridge

Galperin H, Mariscal J (2007) Mobile opportunities: poverty and telephony access in Latin America and the Caribbean. Available at http://dirsi.net/files/REGIONAL_FINAL.Pdf DIRSI/IDRC Accessed 5 January 2015

Gerschenkron A (1962) Economic backwardness in historical perspective. Praeger, New York

Gillwald A (ed) (2005) Towards an African e-Index: ICT access and usage. Witwatersrand University School of Public and Development Management, The LINK Centre, Johannesburg

Hobday M (1995) Innovation in East Asia. Edward Elgar, Cheltenham

Jack W, Suri T (2010) The economics of M-PESA. Sloan, MIT. Available via http://www.mit.edu/~tavneet/M-PESA.pdf. Accessed 12 August 2012

James J (2009) Leapfrogging in mobile telephony: a measure for comparing country performance. Tech Forecast Social Change 76(7):991–998

James J (2011) Sharing mobile phones in developing countries: implications for the digital divide. Tech Forecast Social Change 78(4):729–735

James J (2013) Digital interactions in developing countries: an economic perspective. Routledge, Oxford and New York

Mas I, Radcliffe D (2010) Mobile payments go viral: M-PESA in Kenya. Capco Inst J Financ Transform 32:169–182

Minges M (2012) Key trends in the development of the mobile sector. In: World Bank, Maximizing Mobile, Washington DC

Porritt J (2005) Capitalism as if the world matters. Earthscan, London

Rashid A, Elder L (2009) Mobile phones and development: an analysis of IDRC-supported projects. EJISDC 36(2):1–16

Ray D (1998) Development economics. Princeton University Press, Princeton

Research ICT Africa (RIA) (2008) ICT access and usage in Africa. Policy paper vol 1, paper 2, Cape Town, South Africa

Richardson D, Ramirez R, Haq M (2000) Grameen telecom's village phone programme in rural Bangladesh: a multi-media case study. Available at http://www.microfinancegateway.org. Accessed 3 Sept 2005

Samuel J, Shah N, Hadingham W (2005) Mobile communications in South Africa, Tanzania and Egypt: results from community and business surveys. Vodafone policy paper series, no. 2:44–52

Sen A (1985) Commodities and capabilities. North-Holland, Amsterdam

Stewart F (1977) Technology and underdevelopment. Macmillan, London

The Economist (2008) The limits of leapfrogging. Available via http://www.economist.com/node/10650775. Accessed 13 May 2011

Waverman L, Meschi M, Fuss M (2005) The impact of telecoms on economic growth in developing countries. Vodafone policy paper series, no. 2:10–24

World Bank (2012) Maximizing mobile. Washington DC

Chapter 5
Micro, Macro and Scaling-Up Effects

Abstract This chapter covers much of the work that has been done on the economic impact of mobile phones in developing countries. It covers micro, macro and the transition that is sometimes made from the former to the latter level (that is, scaling-up). Research at the micro level is relatively plentiful and is concerned mainly with the improved communication and information in various markets, that mobile technology can help to bring about. Unfortunately, however, few of the studies at this level specifically refer to the poor and there is consequently a rather glaring research gap to be filled. There are, of course, exceptions to this general pattern and these tend, encouragingly, to reveal something of a bias in favour of the poor (such as for example the greater benefits that accrue to this group in the case of Village Pay Phones (VPPs) in Bangladesh. Similarly, at the macro level, three out of the four studies that were identified find a greater effect of mobile phones on growth, the lower is the presence of fixed-lines in a country. This favours the poorer countries, where the older technology tends to be at least in evidence. But here, as at the micro level, there is too little evidence to reach any firm conclusions. The final part of the chapter deals with three cases of successful scaling up of projects that began on a small scale. Yet, in this area too, scant attention is paid to the impact of the process on the poor, suggesting still another area for further research.

Keywords Pro-poor · Markets · Fixed-phones · Leapfrogging · M-PESA · Grameen

I have already taken note of the varying amounts of consumer surplus that accrued to different income groups in the Village Phone Project in Bangladesh. However, this is certainly not the only micro effect nor the only market. Similarly, I have referred to one study at the macro level by Waverman et al. (2005), but there are others in this category that ought to be considered. Finally, there has thus far been

© The Author(s) 2016
J. James, *The Impact of Mobile Phones on Poverty and Inequality in Developing Countries*, SpringerBriefs in Economics,
DOI 10.1007/978-3-319-27368-6_5

no mention of the scaling up of pro-poor innovations to national levels. But this phenomenon needs to be discussed because it may have substantial welfare effects on the poor. The purpose of this chapter, accordingly, is to overcome these various limitations and in so doing, to provide a somewhat fuller version of the impact of mobile phones on the poor. It should be noted at the outset, though, that in none of the areas concerned—micro, macro and scaling-up—have I found more than a handful of studies that bear directly on the poor (individuals or countries). It is to be hoped, nonetheless, that the evidence gleaned from these studies will be sufficient to point, more or less tentatively, to certain tendencies and topics for further research. I will try moreover, to relate the findings from this chapter to what was discovered in Chaps. 3 and 4 regarding the pro-poor effects of mobile phones in developing countries.

I turn first to a discussion of the micro-economic effects.

5.1 Micro-economic Effects

Papers on the micro-economic effects of mobile phones tend to focus heavily on markets and their imperfections. One such study is by Jensen (2007) who rightly points out that the major theorems of welfare economics—such as the law of one price—depend crucially on the assumption of perfect information in the relevant markets.

This assumption, however, could hardly have been more inaccurate in the case of Kerala's traditional fishing industry, characterized as it was by large spatial price differences across only a relatively narrow range of coastal villages. Jensen then poses the question of whether information imperfections could be ameliorated or eliminated by the introduction of mobile phones in Kerala's fishing industry. He is able to examine this question by drawing on data before and after the intervention. In this way he can compare a situation which was riddled with all manner of imperfections with one that came close to the welfare ideal of perfect competition (see also Aker 2010).

One of the most severe information imperfections that prevailed in the period before mobile telephony was that 'producers and traders' often had very limited information about prices in markets other than those in nearby villages. The result was that they (the villagers) usually sold their catch in those neighbouring locations, instead of in other places where prices were higher. In a rare reference to poverty in developing countries, Jensen notes that the functioning of output markets 'plays a central role in determining the incomes of the significant fraction of households engaged in agriculture, forestry, or fisheries production in low-income countries; for most of the world's poorest, living standards are determined largely by how much they get for their output' (Jensen 2007, p. 880).

After mobile phones had been widely dispersed to fishermen along the Kerala coast, the market for fish in the area changed quite dramatically.[1] In particular, Jensen finds that:

> Price dispersion was dramatically reduced with the introduction of mobile phones; the mean coefficient of variation of price across markets ...... declined from 60-70 to 15 percent or less. In addition, there were almost no violations of the Law of One Price ..... Further, waste, averaging 5-8 percent of daily catch before mobile phones was completely eliminated. Overall, the fisheries sector was transformed from a collection of essentially autarkic fishing markets to a state of nearly perfect spatial arbitrage (Jensen 2007, p. 883).

More generally, as other authors have pointed out, micro and small enterprises in developing countries often are run by, or employ, poor persons whose incomes can be increased through mobile phones. This can occur, for example, by means of increased productivity, improved supply-chain management, or better methods of organization. Perhaps the main point is that for low-income firms, no less than for poor consumers, there is no real alternative to the mobile phone.

Another market which typically suffers from acute information imperfections is that for labour. It is often the case for example, that there is not only poor knowledge of available jobs on the part of those seeking work, but also symmetrically imperfect information on the part of employers about those seeking particular jobs. Against the background of rural South Africa with very little in the way of fixed-line phones, Klonner and Nolen (2008) regard mobile alternatives 'primarily as an improvement in information flows. This has two major effects on the labor market. First, improved spatial integration of the market for wage labor and, second, better conditions for business startups through reduced fixed costs, lower cost of information, outreach to a broader customer base etc. The first effect will in general result in higher wage-employment, while the second effect will result in more business startups and thus more self-employment' (Klonner and Nolen 2008, pp. 4–5).

What the authors find in the empirical part of their paper is relevant in this context insofar as it bears on employment and poverty, the latter being divided into moderate and extreme forms. In a rigorous regression analysis, the outcome is that the 'extreme form of poverty is relieved substantially in all specifications, while moderate poverty ... shows no significant effect in any direction' (Klonner and Nolen 2008, p. 17). Quite why there is a different result for different intensities of poverty is not clear but Klonner and Nolen nonetheless summarize their findings by arguing that 'income growth triggered by cellphone infrastructure benefits the very poor significantly and we may thus understand the process of cellphone expansion as pro-poor' (Klonner and Nolen 2008, p. 17).

These authors are, in fact, one of the very few to have examined the impact of mobiles on poverty at the micro level in developing countries. Another is the group of authors (Bayes et al. 1999) who, as noted above, studied the consumer surplus of poor users of the Grameen Village Pay Phone Project (discussed also later in

---

[1]Jensen (2007) notes that CS rose by 6 %. Recently, however, Hausman and Liu (2014) have significantly increased that estimate.

this chapter). Then there is the collection of essays contained in the recent publication of a book entitled 'ICT Pathways to Poverty Reduction: Empirical Evidence from East and Southern Africa', edited by Adera et al. (2014).

In spite of its highly promising title, however, this volume does not in fact throw much light on the impact of mobile phones on poverty in the regions concerned. This is not because the book fails to live up to its title, but rather that it tends to conflate mobiles with other forms of ICT such as the internet. Thus, it becomes difficult to disentangle the separate effects of each of this group of technologies.

One chapter in the volume, however, by Mascarenhas (2014), does present a clearer—albeit somewhat approximate—picture of the specific impact of mobile phones on the poor in Tanzania. Using two data points, she compares the poverty situation before and after the introduction of mobile phones in a control and a benefit group. (Mobiles were not the only ICT to be introduced to the benefit group but they did constitute the dominant share of the increased expenditure). Therefore, there was in fact no deliberate attempt to separate out the mobile phone from other forms of ICT. The ability to isolate the impact of the former technology is purely fortuitous, resulting from its dominance in increased expenditure on the latter. The results are thus only an approximation to the 'true' contribution of mobile phones on poverty. They do, nonetheless, provide a clear picture of the impact on the poor. That is, that over a decade, the gains resulting from ICT for the most poor are twice that for the non-poor (Mascarenhas 2014, p. 234).

Another chapter, by Agüero et al. (2014), does explicitly consider only the impact of mobile phones and it carefully describes the mechanisms through which these technologies benefit the group of respondents. Unfortunately, however, it does not divide the sample according to income and does not therefore provide evidence of the extent to which mobile phones impact on poor individuals and firms. This is a distinct anomaly given the title of the book in which the chapter appears.

All in all, it seems fair to say that the discussion of conceptual relationships between mobile phones and poverty in developing countries far exceeds those of an empirical nature. This is a gap that should soon be filled, not least because there are important questions about the extent to which some of the groups living in poverty actually benefit from a reduction in information imperfections. Perhaps the most serious of those questions has to do with the fact that there are many millions of poor people who are not (or are barely) connected to markets, such as those who are disconnected from markets and who as a result may be less prone to reaping the benefits from improvements therein.

Minot and Hill (2007) pose the question of why small farmers in developing countries 'produce largely for their own consumption' in spite of the benefits that markets provide. One reason has to do with the factors that raise marketing costs and the other concerns the risks entailed in producing for the market. As to the former, one can point to unfavourable infrastructural conditions such as poor roads, high travel costs and a lack of market information. Minot and Hill (2007, p. 1) note for example that 'studies of Laos, Malawi and Zaire, among other countries, [which] have found that the level of commercialization declines with the distance from roads and markets. Poor farmers have only small amounts to sell, making long-distance travel to sell their products unprofitable.' There is some scope here for mobile phones to reduce the marketing costs associated with production for the market but small farmers often tend to be among the poorest of the poor and may accordingly be unable to afford a handset and the price of calls (unless, of course, they rely on one of the many rental markets that have been described in previous chapters).

As has just been pointed out, the second reason given by the authors for favouring subsistence production over the market has to do with the risks of commercialization. For example, 'producing for markets sometimes requires intensive and costly input use, which results in substantial risk for small farmers when yields are uncertain' (Minot and Hill 2007, p. 1). Whether this or other forms of risk associated with commercialization can be overcome by mobile phones is a matter for future research (indeed, I would venture to suggest that the entire effect of these products on the poorest segments of society is an area that warrants close scrutiny).

Note finally that even if the nature of rural poverty limits the gains from improvements in information through mobile phones, there are still important benefits to be reaped at the micro level from non-economic sources. I have already referred for example to benefits associated with mobile telephony from increased safety, improved healthcare and so on.

5.1.1 Macro-economic Effects

Macro-economic studies of the impact of mobile phones on growth (and hence hopefully on poverty) are unfortunately no more plentiful than the micro ones I have just reviewed. Qiang (2009a), for example, identifies just four such cases. These are summarized in Table 5.1.

Note first that in each of the four cases, there is a strong impact of mobile phones on growth in developing countries. This will generally tend to reduce poverty but as shown below, much depends on the way incremental output is distributed among different income groups. The second point to note is that most of the studies (with the exception of Sridhar and Sridhar) support the claim that the growth dividend is higher in poor than in rich countries. This much is obvious from the studies by Waverman and the World Bank, which explicitly calculated the difference in growth rates between the two groups of countries and it can also be inferred from the Lee exercise, if one equates countries with low rates of fixed-line adoption with relatively poor developing countries. For it is just such countries which, according to Lee, exhibit a relatively powerful growth response to mobile phone adoption. The reason, as suggested in Chaps. 3 and 4, is that in the absence of fixed-lines and basic infrastructure, there are very few alternatives to the mobile phone, thus enhancing its impact on growth.

Table 5.1 Studies on the macro-economic impact of mobile phones

Author(s)	Country coverage	Results (Qiang 2009a)
Sridhar and Sridhar (2004)	63 developing countries	'We find significant effects of ... cell phone penetration on economic growth, but lower than that found for OECD countries, dispelling the convergence hypothesis' (p. 25)
Lee et al. (2009)	44 Sub-Saharan countries	'We find that mobile cellular phone expansion is an important determinant of the rate of economic growth in Sub-Saharan Africa. Moreover, we find that the marginal impact of mobile tele-communication is even greater wherever land-line phones are rare' (p. 1)
Waverman et al. (2005)	92 high and low income countries	'The growth divided of increasing mobile phone penetration in developing countries is therefore substantial' (p. 11). Mobile phones have twice as large an impact on developing as developed countries
Qiang (2009b)	120 countries	'For every 10 percentage point increase in the penetration of mobile phones, there is an increase in economic growth of 0.81 percentage points in developing countries versus 0.60 percentage points in developed countries' (p. 8)

The next question is, of course, whether and to what extent the growth impact of this technology actually benefits the poor in developing countries. What occurred in China and India describes much of the developing country experience as a whole since these countries are so large. The rapid economic growth there 'contributed to the world's poverty reduction in a major way' (Helpman 2004, p. 107). To be precise, between 1980 and 1992 China's real income grew at an average yearly rate of 3.58 % and India's increased at the slightly lower figure of 3.12 %. The associated declines in the number of poor people over the same period were, respectively from 360–530 million to 110–166 million. Clearly, 'the economic growth of China and India was associated with massive poverty reductions' (Helpman 2004, p. 108).

Nor, according to Dollar and Kraay (2002), was this an untypical result for the rest of the developing world. For, using a large number of countries, 'they showed that the average real income per capita of a country's poorest quintile moved practically one-to-one with the average real income per capita of the country's entire population. This relationship is ... very tight' (Helpman 2004, p. 108). In other words, the poorest quintile benefits as much from growth as the average person in the countries concerned.

Several considerations need to be borne in mind when interpreting these results. The first is that change (dynamics) can never be adequately captured by cross-section research. This was the problem for example with (cross-sectional) research designed to test the Kuznets hypothesis that at low levels of income, growth creates inequality. Following Helpman (2004) it seems safer to conclude that 'on average growth has raised the income of the poor round the world' (p. 109).

The second consideration that needs to be borne in mind concerns variations around the average levels of poverty change. For the average may conceal marked variations in outcomes, such as, for example, cases of a very low contribution of growth to poverty reduction.[2] It therefore seems necessary to identify the factors that are conducive to pro-poor growth and those that are not.

Fortunately, in a paper entitled 'Does Economic Growth Reduce Poverty?' such a task has been undertaken by Roemer and Gugerty (1997). They divide the effects of different policy stances into two categories: one is the extent of their impact on growth and the other is the translation of that growth into income gains for the poorest groups. The most favourable outcomes are those where both effects are (strongly) positive and the least favourable combine limited (or even negative) effects on growth with a weak poverty response to that growth (assuming of course that it is positive).

Resource abundance and openness tend to have opposite effects on poverty primarily because of their differing impacts on growth (weak and strong, respectively). The authors stress, however, that the weak or negative effects of resource abundance—arising partly from an inegalitarian effect of growth on distribution—can be overcome, partly or fully, by the pursuit of policies that are explicitly pro-poor. I am referring here, for example, to policies of land reform or support to small-scale farmers.[3] Note in this regard that China and certain successful East Asian countries such as South Korea, exhibited not only a significant dependence on exports but also a preponderance of labour-intensive products among those exports, increasing the share of unskilled workers in the labour force and reducing the extent of national poverty.[4]

5.2 Scaling-Up Effects

The macro studies in the previous section were all based on a simulation methodology. That is, a simulated estimate of what would happen to economic growth if the number of mobile phones was to increase by a certain percentage in a given country or countries. The process of scaling-up on the other hand, involves

---

[2]One such case, although it is not a developing country, is the United States from the mid-1970s to the mid-1990s (Helpman 2004).

[3]For a list of alternatives see Griffin and James (1981).

[4]For a full discussion of the East Asian export experience and its reliance on labour-intensive methods of production, see James (1987).

observing how an initially small and successful mobile phone intervention spreads so widely as to have a discernible impact at the macro level. In other words, it is concerned with the transition of a mobile project or programme from the micro to the macro level of an economy.

As such, scaling-up is not a new concept. Indeed, it has been widely discussed in the development literature on aid effectiveness during the 1970s and 1980s. In that context 'A key constraint that needs to be overcome is that development interventions … are all too often like small pebbles thrown into a big pond: they are limited in scale, short lived, and therefore without lasting impact. This may explain why so many studies have found that external aid has had only a weak or no development impact in the aggregate at the global and at the country level' (Hartmann and Linn 2008, p. 2).

Unfortunately, much of the same can be said of the more recent experience with mobile phones in developing countries. In Africa, for example, 'the development and testing of mobile applications have not been the greatest challenge faced by the technology community … Rather, it is the scaling of the applications, … that has been noted as the greatest challenge' (Omwansa et al. 2013, p. 3).

Against this background, I now turn to examine the successful scaling up of three cases that were briefly mentioned above, namely, South Africa's community phone shops, Grameen Village Phones, and M-PESA.[5]

5.2.1 Three Cases of Scaling-Up: Community Phone Shops, Grameen Telecom and M-PESA

In the case of South Africa's Community Phone shops, Vodacom enters into a franchise agreement with individual entrepreneurs, who operate mostly out of informal (or 'spaza') shops. From the outset, the state was heavily involved in the goal of reaching low-income consumers in disadvantaged areas. Although Vodacom had as its main business the meeting of demand by upper and middle-income groups, the community phone shops were entrusted with the task of serving the low-income end of the market. And while the company's main task was pursued in a relatively unfettered environment, serving the poor was closely regulated by institutions of the state.

For example, Vodacom's initial licensing agreement required the company to provide 22,000 public mobile phones over a five-year period in isolated and underserved areas of the country. Moreover, prices were set at less than one-third of their usual levels and the individual entrepreneurs were allowed to retain one-third of the calling profits from their phone shops (which, more often than not, were located in used shipping containers).

---

[5]Hartmann and Linn (2008) describe a number of successful scaling-up projects from the development literature. Unfortunately, the explanations of scaling-up in mobile phones make no attempt to forge links with this more general discussion.

Part of the success of the model lies in its exploitation of rental markets rather than the sale of individual handsets. Indeed, according to Reck and Wood (2003, p. 7) 'The shared-access model providing telephone connectivity for entire communities at an affordable cost per use, is quite profitable'. But what is described as a 'significant factor' in the success of the Vodacom case lies, according to these authors, 'with the existing infrastructure that was built to service Vodacom's traditional cellular customers' (p. 16). These customers in effect 'subsidize the services provided for Community Services customers. Without the traditional services, there would be no infrastructure and Community Services simply would not be able to justify the cost to develop its own infrastructure. Community Services is successful because of its ability to utilize an existing resource without incurring additional costs' (Reck and Wood 2003, p. 16).

This indeed is a recurring theme in the two following cases as well.

The Village-Pay-Phone (VPP) project in Bangladesh embodies some of the same features as the previous case; for example, it is a shared-access model which makes mobile phones available at the community level, by selling air-time rather than individual ownership of the technology. The VPP project is made possible by the support of two major Bangladeshi institutions, the well-known Grameen Bank and Grameen Phone (see below). In this way, from its very inception the project was well embedded in the institutional infrastructure of the country (as was also the case with the Vodacom Phone Shops in the previous example).

The Grameen Bank took upon itself the task of making credit available to poor village women ('phone ladies') entrepreneurs, for the purpose of purchasing a mobile phone from Grameen Telecom, an NGO, which adopted the view that 'even if rural villagers in Bangladesh could not afford a phone individually, they could afford one collectively' (Lawson and Meyenn 2000, p. 1). The last institution in this story is Grameen Phone, a mobile operator, which sells airtime to Grameen Telecom for resale in mainly rural areas and which subsequently came to dominate the market in that country.

The relationships between these various relationships are, according to Cohen (2001), what sustains the VPP initiative. Thus,

> Grameen Telecom's village phone venture as structured in Bangladesh would not be feasible without access to the credit and bill collection services provided by Grameen Bank and the infrastructure and urban mobile phone network provided by Grameen Phone. Village phones would be far less successful if GP were not able to discount by 50 % the rate charged to GT for a phone call, an underlying subsidy made possible by a transfer of profits from the more profitable urban part of the business to the rural sector (Cohen 2001, p. 3).

In this last respect, there is a close similarity to the Vodacom model in South Africa. One should not forget, either, that, as noted in a previous chapter, the consumer-surplus reaped by the users of the Village Phones was highest among the poor, giving rise to a major boost in the replication of the project around the villages in the country. Indeed, according to an estimate provided by the Grameen organization itself, as many as 67,000 villages have been provided with a phone, providing access to some 67 million people. This is nothing, if not a highly successful example of scaling up a mobile project, aimed largely at the rural poor.

It is worth emphasising here that the two cases of scaling-up that have been described thus far, will have had certain macro effects on the poor beyond the gains from use of the phones themselves. For example, in both cases meaningful gains were seen to accrue to the many individual entrepreneurs who actually operated the phones, particularly the independent franchises in the Village Community phone project and the 'phone ladies' in the Bangladesh case. Their increased spending in turn may well have prompted further macro-economic effects with feedback loops to others among the poor.

No less impressive than these two cases, however, and possibly more so, is the M-PESA mobile banking operation in Kenya, which is used by some 17 million of that country's population. Indeed, led by Safaricom—a subsidiary of Vodacom— M-PESA can now rightly be described as the world leader in mobile money. I have no evidence, though, of the extent to which the poorest income groups in that country actually benefit from the scheme, but one prominent academic institution reports 'outstanding rates [of adoption] among low-income customers' (Brookings Institution 2013).

Although the scaling up of the Kenyan scheme was heavily reliant on the market mechanism, this mechanism had to be carefully managed because in the context of the spread of mobile money, the market is very much a double-edged sword. The reason is basically that it does not itself favour incremental change in the process of scaling up. If left to itself, the system may indeed get stuck in what economists term a 'low-level equilibrium trap'. Mas and Radcliffe (2011) provide three reasons for this possibility. The first concerns network effects as these were described above. While these 'can help a scheme gain momentum once it reaches a critical mass of customers, they can make it difficult to attract early adopters when there are few users on it' (Mas and Radcliffe 2011, p. 7). The second reason is labelled 'the chicken-and-egg trap'. On the one hand, that is, it is difficult to make the product attractive to buyers while there are relatively few sellers, and on the other hand, the converse is also true. Finally there is the question of trust. Buyers of mobile banking services have to give more of themselves with mobile money than is true of simply renting phone time (as is true of the two previous examples). In particular, 'Customers have to become comfortable going to non-bank retail outlets to meet their cash-in/out needs and initiating transactions through their mobile phones' (Mas and Radcliffe 2011, p. 7).

From one particular point of view, therefore, the story of M-PESA needs to be understood in terms of how the programme managed to escape from these negative impulses (when most of the other mobile phone schemes in developing countries appear to be caught precisely in the low-level trap described above).[6]

More specifically, the task becomes one of understanding how the company was able to reach the 'tipping point', after which the underlying forces became cumulatively favourable or 'self-propagating'.

---

[6]Economists such as Sachs (2006) believe that much of Africa is stuck in such a trap, but his view has been subject to harsh criticism.

Stripped to its essentials, the answer to this question is that,

Safaricom didn't grow M-PESA through tepid, incremental steps because, when it comes to mobile money, gradualism likely leads to failure. After small pilots involving less than 500 customers, Safaricom launched M-PESA nationwide, making heavy up-front investments so it could reach a critical mass of customers in a short time-frame. Many deployments have potential to scale, but are stuck in the 'sub-scale trap' because their promoters either under-estimate the investments needed to achieve scale, or are unable or reluctant to make these investments because they can point to only one major success – M-PESA (Mas and Radcliffe 2010).

There were of course other reasons for the as yet unmatched degree of scaling up achieved by Safaricom and M-PESA to which the interested reader is referred.[7]

The World Bank (2012) provides three of them. The first was the supportive role played by regulators. Most notably, the Central Bank permitted 'regulation to follow innovation', while at the same time convincing the market of its serious 'oversight' of the project. It was decided, for example, that the mobile money agents were not to be stringently regulated, since they were not in fact offering banking services. The second reason has to do with the role of the phone operator, Safaricom, which, even by 2007, had captured more than half the market share. This was an important asset since it went with a larger network of airtime resellers who could be converted into mobile money agents.[8] The final reason is that the firm's management stressed the importance of managing people rather than technology. Indeed, as the World Bank sees it, 'The true secret of M-PESA's success is the management of the agent network, which grew from 300 initially to almost 30,000 today' (World Bank 2012).

Insightful though it often is, however, the literature related to these three cases suffers a similar weakness as that already pointed out in the case of micro and macro effects, namely, that there is relatively little on the poverty impact of the processes described. As in those cases, therefore, there is a strong argument for research designed specifically to rectify the problem.

5.3   Conclusions

The purpose of this chapter was to approach the impact of mobile phones on poverty through three different, but related, conceptual lenses; micro and macro-economics and scaling up pro-poor innovations to the national level. In this endeavour, there were several points of overlap with Chaps. 3 and 4 but almost all of the discussion was based on a different set of literature.

---

[7]There is already quite a sizeable literature on the reasons for M-PESA's success. See for example Jack and Suri (2010), Mas and Radcliffe (2010, 2011), The Economist (2010, 2013), World Bank (2012).

[8]This is where there is again a reliance on existing resources as in the other two cases. The airtime resellers already in the Safaricom network could be converted into mobile money agents, thus avoiding having to start the latter from scratch.

At the micro level I found that the conceptual analysis of mobile phones and poverty has outrun the empirical work. There were very few studies that threw any econometric light on this question. Those studies that do exist, find that the effect tends to be pro-poor, but much research remains to be done.

The macro studies focus on the impact of mobile phones and growth (and hence hopefully on poverty), but they are unfortunately also limited in number. The results, moreover, were less than unanimous: some of the cases found the predicted pro-poor effect on different countries, but another found the opposite to be true (all the studies however find a positive effect of mobiles on growth).

The final section—on upgrading a successful micro project to the national level—was cast against a background of largely unsuccessful cases in developing countries. Apparently, the process is very complex and the few successful cases are difficult to emulate. One has only to consider here the attempt to replicate the mobile-banking system called M-PESA in Kenya.[9] Though there is a presumption that all the exceptional cases considered were of benefit to the poor, there is very little hard evidence to prove it (apart, that is, from the CS that was shown in the Grameen phone case, to be higher for this group than the non-poor). It would be desirable for researchers to supply the missing evidence and to connect up with the voluminous literature on scaling up micro-projects in general (see for example, Hartmann and Linn 2008).

What evidence is available in this chapter, is broadly supportive of what was found in Chaps. 3 and 4 regarding the pro-poor effects of mobile phones but the results here are not numerous or (on occasion) not rigorous enough to offer any firm conclusion.

References

Adera E, Waema T, May J, Mascarenhas O, Diga K (eds) (2014) ICT pathways to poverty reduction. Policy Action, Rugby

Agüero A, Barrantes R, Waema T (2014) Livelihood and ICTs in East Africa. In: Adera E et al (eds) ICT pathways to poverty reduction. Policy Action, Rugby

Aker J (2010) Information from markets near and far: mobile phones and agricultural markets in Niger. Am Ec J Appl Econ 2:46–59

Bayes A, von Braun J, Akhter R (1999) Village pay phones and poverty reduction: insights from a Grameen Bank initiative in Bangladesh. Center for Development Research, Bonn. Available via http://www.zef.de/uploads/tx_zefportal/Publications/zef_dp8-99.pdf. Accessed 23 Jan 2001

Brookings Institution (2013) Mobile-money: a technological game changer for tackling global poverty. Available at http://www.brookings.Edu/blogs/up-front/posts/2013/03/12-mobile-money-chandy. Accessed 12 May 2014

Cohen N (2001) What works: Grameen telecom's village phones. World Resources Institute, Washington DC. Accessed 3 July 2014

Dollar D, Kraay A (2002) Growth is good for the poor. J Econ Growth 7:195–225

Griffin K, James J (1981) The transition to egalitarian development. Macmillan, London

---

[9]These attempts are occurring, for example, in the Philippines and Tanzania.

Hartmann A, Linn J (2008) Scaling up—a framework and lessons for development effectiveness from literature and practice. Brookings. Available at http://brookings.edu/research/papers/2008/10/scaling-up-aid-linn. Accessed 22 Feb 2014

Hausman J, Liu Z (2014) Mobile phones in developing countries. Working paper, Consortium on financial systems and poverty. Available at http://economics.yale.edu/sites/default/files/hausman-07-oct-2014.pdf. Accessed 24 Mar 2015

Helpman E (2004) The mystery of economic growth. Cambridge, Belknap

Jack W, Suri T (2010) The economics of M-PESA. Available at http://www.mit.edu/~tavneet/M-PESA.pdf. Accessed 14 Aug 2012

James J (1987) Population and technical change in the manufacturing sector of developing countries. In: Johnson D, Lee R (eds) Population growth and economic development. University of Wisconsin Press, Wisconsin

Jensen R (2007) The digital provide: information (technology), market performance, and welfare in the South Indian fisheries sector. QJE 122(3):879–924

Klonner S, Nolen P (2008) Does ICT benefit the poor? Evidence from South Africa. Available via http://privatewww.essex.ac.uk/~pjnolen/KlonnerNolenCellPhonesSouthAfrica.pdf. Accessed 12 Dec 2014

Lawson C, Meyenn N (2000) Bringing cellular phone service to rural areas. Public Policy for the Private Sector, Note No. 205

Lee S, Levendis J, Gutierrez L (2009) Telecommunications and economic growth: an empirical analysis of Sub-Saharan Africa. Available at http://papers.ssrn.com/so13/papers.cfm?abstract_id=1567703. Accessed 9 Nov 2012

Mas I, Radcliffe D (2010) Mobile payments go viral: M-PESA in Kenya. SSRN. Available at http://papers.ssrn.com/sol3/papers.cfm?abstract_id=1593388. Accessed 29 Oct 2013

Mas I, Radcliffe D (2011) Sealing mobile money. J Pay Strategy 5(3):1–19

Mascarenhas O (2014) Conclusion and recommendations. In: Adera E et al (eds) ICT pathways to poverty reduction. Policy Action, Rugby

Minot N, Hill R (2007) Developing and connecting markets for poor farmers. 2020 Focus brief on the world's poor and hungry people. Available via http://lib.icimod.org/record/13063/files/4324.pdf. Accessed 15 Mar 2014

Omwansa T, Crandall A, Waema T (2013) The gap between mobile application developers and poor consumers: lessons from Kenya. In: CPR South/CPR Africa conference. Available at http://papers.ssrn.com/sol3/papers.cfm?abstract_id=2364263. Accessed 3 Jan 2014

Qiang C (2009a) Mobile telephony: a transformational tool for growth & development. Private Sect Dev 4:7–16

Qiang C (2009b) Telecommunications and economic growth. Unpublished paper, World Bank

Reck J, Wood B (2003) What works? Vodacom's community services phone shops. World Resources Institute, Washington DC. Available at http://pdf.wri.orgldd_vodacom.pdf. Accessed 18 Apr 2010

Roemer M, Gugerty M (1997) Does economic growth reduce poverty? CAER discussion paper No. 4. Harvard Institute of International Development. Available via http://pdf.usaid.gov/pdf_docs/pnaca655.pdf. Accessed 27 Oct 2007

Sachs J (2006) The end of poverty. Penguin, New York

Sridhar K, Sridhar V (2004) Telecommunications infrastructure and economic growth: evidence from developing countries. Working paper, Indian National Institute of Public Finance & Poverty. Available via http://econpapers.repec.org/paper/npfwpaper/04_2f14.htm. Accessed 19 Mar 2007

The Economist (2010) Out of thin air. Available at http://www.economist.com/node/16319635/print. Accessed 9 May 2011

The Economist (2013) Why does Kenya lead the world in mobile money? Available at www.economist.com/blogs/economist-explains/2013/05/economist-explains-18. Accessed 1 Jan 2014

Waverman L, Meschi M, Fuss M (2005) The impact of telecoms on growth. In: The Vodafone policy paper series, no. 2. Available via http://www.vodafone.com/content/dam/vodafone/about/public_policy/policy_papers/public_policy_series_2.pdf. Accessed 15 July 2008

World Bank (2012) How Kenya became a world leader for mobile money. African can end poverty. Available at http://blogs.worldbank.org/africacan/how-kenya-became. Accessed 19 Apr 2013

Chapter 6
Patterns of Mobile Phone Use in Africa

Abstract The distinctive features of this chapter are that it is conducted from an inter-country perspective; that it takes use rather than adoption as the measure of welfare; and that it deals with numerous impact mechanisms other than the purely economic (such as health and safety). Yet, the chapter retains a close affinity with Chaps. 5 and 7; the former because of its main hypothesis that mobile technology will be most widely used in countries lacking a viable alternative to the new technology. The latter chapter because it also challenges the conventional measure of the digital divide which is based on adoption rather than use. Particular attention is paid to the relatively poor East African countries in the sample, which tend to make the most intensive use of mobile phones in the areas of economics, health, social capital and safety. Because it turns out to be the most important of these use mechanisms and since it is not widely discussed in the literature, safety is studied more intensively in the Appendix to this chapter. A key question in this regard is why safety constitutes such a popular mechanism for mobile phone use in the countries concerned. My tentative answer is that it has much to do with the interactions between crime, poverty and inequality.

Keywords Infrastructure · Digital divide · East Africa · Safety · Technology use · Crime

In an article on mobile phones published in 2007, James and Versteeg pointed out that from a welfare point of view what matters is the use of this technology and the benefits it conveys thereby. Yet, as these authors also made clear, the evidence on this topic tends to be concerned with owners and subscribers and it is not helpful to answering questions about how many people benefit from the use of mobile phones and in which specific ways. It is true that there are a few limited studies on this question, such as by Goodman (2005) and Samuel et al. (2005) on the economic and social benefits of mobiles in a select few African countries.[1]

---

[1] I am grateful to Research ICT Africa for making the date available to me. The underlying methodology is contained in Research ICT Africa. net. 'Household, Small Business and Public Institutional e-Access and Usage Survey 2011'.

© The Author(s) 2016
J. James, *The Impact of Mobile Phones on Poverty and Inequality in Developing Countries*, SpringerBriefs in Economics,
DOI 10.1007/978-3-319-27368-6_6

Not until quite recently however has a far more extensive set of survey data been collected, which covers a relatively large number of welfare-enhancing mechanisms and as many as 11 comparable African countries. It is fair to say that the data-set as a whole comprises one of the most comprehensive collections of evidence on mobile phone use now available in developing countries. The purpose of what follows is to examine the said data and especially the welfare patterns that underlie them, where welfare is defined more broadly than only in economic terms.

For this purpose I first divide the data on mechanisms of use into 4 categories, dealing, respectively, with economics, health, social capital and safety. Each of the categories is analyzed initially in the form of a table, which exhibits the countries as columns and the mechanisms as rows. Some countries will of course do better than others and certain mechanisms will contribute more to welfare than others in a table. Ideally, I would be able to explain all the patterns that emerge but more realistically many of them will remain as questions for future research (see conclusions).

In the next phase of the analysis the scope of the chapter is extended to allow for comparisons across tables. To the best of my knowledge this type of cross-tabular calculation does not appear elsewhere in the existing literature. It can address questions such as whether the same countries (or countries from the same region) tend to perform well or poorly across categories or whether the results have a more random character. Here too though some of the observations will generate a research agenda rather than well-formulated questions. As it happens, much of this agenda will need to revolve around finding explanations for the systematically favourable performance of the relatively impoverished countries from the East Africa region (the impact of mobile phones on poverty thus refers here to poor countries, rather than poor people).

But it will also involve the development of an analytical framework in which the use of products and technologies helps to determine their overall impact on welfare. Sen's concept of functionings noted above already captures the essence of such an idea.

6.1 Survey Method and Characteristics of Respondents

As described in the reference given in footnote 1 in Chap 1, the methods used to collect the data accord very well with the goals of this study as these were described above. For one thing, the method includes a diverse group of 11 African countries, some very poor and others among the richest on the continent. There is also a dispersion of the sample countries by region (including East, West and Southern Africa). Within each country, moreover, the survey relies on a relatively large number of respondents as described in the reference contained in footnote 1 in Chap 1. Relatedly, the data are collected with the aid of national statistical methods which help to provide nationally representative information.

(I am thinking here for example of national census sample frames and enumerator areas). Finally, the survey covers an impressively wide range of ways in which mobile phones are actually used to enhance welfare in the relevant countries. For analytical convenience I have divided these mechanisms into 4 categories in the presentation of the results (namely, economic, health, social and safety-related measures). Thus collected, the results are surely one of the most comprehensive now available on the use of mobile phones in developing countries.

Table 6.1 well reflects the basic conditions of mobile phone use in the poorest developing countries since most of the sample belongs to what the World Bank refers to as the low-income category (see the bottom row for the per capita income levels of the eleven sample countries). For example, only in slightly more than half of them does the population own this technology. Among owners only about one quarter are able to use the phone for browsing the Internet. This is surely due in large part to the relative cost of the products that are required for such a purpose. In fact, lack of affordability is given (in the next set of rows) as the main reason for not buying a mobile phone in general (followed by a lack of electricity). Affordability was already discussed in Chap. 3. Finally, from among those using public phone facilities, almost three quarters rely on mobile phone kiosks as opposed to the quarter of respondents who use fixed-line payphones.

There are, however, some notable exceptions to these average tendencies. On the one hand, one should single out the two relatively developed countries, South Africa and Botswana. I shall focus on the former because it is structurally more similar to developed countries than the latter. That South Africa displays exceptional characteristics can be seen from almost every row in the table. To begin with, this country has the highest rate of mobile phone ownership in the sample, which is due primarily to its relatively high level of per capita income (a factor which is often the main explanatory variable in accounting for rates of diffusion of information technology across developed and developing countries).[2] Because of its exceptionally high rate of mobile phone adoption there is very little sharing of mobile phones in South Africa (as is also the case in countries that are described as high income or developed). Indeed, 7.3 is the lowest percentage of all countries in the sample. Or again, at 51 %, this relatively affluent country exhibits double the average percentage of those who are able to use their mobile phones to browse the Internet. In the following rows, dealing with the inability of respondents to own a phone, the reason is less to do with affordability compared to the other countries and more to do with broken and stolen products (with the latter reflecting perhaps the unusually high crime rate in the country).[3] A final observation concerns the use of fixed-line or mobile phones in public places. Being more advanced than almost all other African countries at the time when mobile phones were introduced on the continent, South Africa had by then built up a larger stock

---

[2]See for example Dewan et al. (2004).

[3]For data see UNODC.org.

Table 6.1 About the respondents (%)

	Uganda	Kenya	Tanzania	Rwanda	Ethiopia	Ghana	Cameroon	Nigeria	Namibia	S.Africa	Botswana	Average
Do you own a mobile phone?	46.7	74.0	35.8	24.4	18.3	59.5	44.5	66.4	56.1	84.2	80.0	53.6
% prepaid	98.0	99.8	99.5	90.1	98.4	97.4	99.0	99.0	91.8	87.5	94.4	95.9
Do you share your mobile with others?												
No	49.5	71.4	78.5	72.7	67.6	77.5	78.0	79.6	53.7	92.7	67.7	71.7
Yes	50.5	28.6	21.5	27.3	32.4	22.5	22.0	20.4	46.3	7.3	32.3	28.3
Is your mobile phone capable of browsing the internet?	14.9	32.3	19.2	19.1	6.5	28.5	14.9	22.7	30.7	51.0	29.5	24.5
Why don't you have a mobile phone?												
– Cannot afford it	76.5	82.3	87.3	81.6	88.3	70.6	69.7	76.6	71.8	62.8	81.1	77.1
– No mobile coverage	30.8	4.2	9.3	12.9	23.3	21.1	42.6	15.0	15.3	1.5	5.8	16.5
– No electricity at home	79.9	39.0	87.6	64.9	50.3	40.2	50.5	56.8	73.2	8.5	48.9	54.5
– No one to call	20.6	4.7	23.2	17.5	24.1	13.4	16.0	17.4	17.7	5.8	7.0	15.2
– Phone broken	9.8	13.3	12.8	4.5	0.1	18.1	5.7	6.9	16.8	28.7	20.3	12.5
– Phone stolen	7.3	20.8	3.8	3.9	0.3	10.4	9.0	13.8	10.1	17.5	13.2	10.0

(continued)

Table 6.1 (continued)

	Uganda	Kenya	Tanzania	Rwanda	Ethiopia	Ghana	Cameroon	Nigeria	Namibia	S. Africa	Botswana	Average
What type of public phones are you using most?												
– Telephone booth (fixed line operator)	18.7	0.0	90.3	5.1	20.2	0.9	4.8	7.6	95.1	35.9	4.4	25.7
– Telephone kiosk (umbrella operator)	81.2	100.0	9.7	93.8	75.5	93.3	94.7	91.7	4.9	62.3	95.6	73.0
GNP per capita	$1300	$1800	$1500	$1300	$1100	$3100	$2300	$2600	$7500	$1100	$16,200	$4300

Source Research ICT Africa

of fixed-line equipment. It is on this stock that the country still depends relatively heavily in the use of public phones, as can be seen in Table 6.1.

On the other hand there are countries that are underdeveloped even by African standards. One might expect these countries to exhibit the opposite tendencies that have just been described in relation to South Africa. Consider for example the case of Ethiopia which is shown in the table as having the lowest per capita income in the sample (and, one suspects, on the continent as a whole). It can be quickly discerned that the expectation for this country is strongly confirmed. In particular, it comes last in the sample with respect to ownership of mobile phones and Internet browsing capability; affordability is a more severe constraint than for any other country and the extent of sharing is among the highest (being the mirror image of a low ownership rate in the country).

At least at the extremes therefore income helps to explain some variation in the data by country (and the more so if Botswana is included in the comparison). A relevant question is whether this variable will continue to exert an influence on the patterns of use shown in the next section. There I shall examine the specific ways in which the benefits of mobile use are actually extracted by members of the eleven sample countries.

6.2 Results for Four Mechanisms of Mobile Phone Use

As noted above I have divided the results into four categories dealing with economics, health, social capital and safety. It should be borne in mind though that for some items the classification into these groups is not entirely clear-cut and other possibilities could be considered.

6.2.1 Economics-Related Mechanisms

What is notable is in the first place that all the top-rated countries in the table are drawn from East Africa. More specifically, 13 out of the 18 cells in the table belong to this region, an outcome that is very unlikely to have been generated by chance. One part of an explanation of these patterns has to do with M-Pesa (see below) and its confinement largely to Kenya, Tanzania and Uganda. This mobile-money venture began in the first-mentioned country and then spread to the other two countries in the region. As shown in Table 6.2, however, mobile money has spread far more widely in Kenya than the other countries. Indeed, the rate of use of m-Pesa in Kenya is almost 80 %, compared to the sample average of just 19 %. It is in fact the former value that largely accounts for the fact that this country does best on average for all the economics related mechanisms of mobile phone use (see last row in Table 6.2).

Table 6.2 Economics-related mechanisms of mobile phone use

	Uganda	Kenya	Tanzania	Rwanda	Ethiopia	Ghana	Cameroon	Nigeria	Namibia	South-Africa	Botswana	Average
% using mobile phone to find work	51.9	38.1	33.6	52.1	42.2	33.3	43.2	29.4	42.7	33.8	38.2	39.9
% with saving on time & travel cost	84.7	88.3	67.7	63.9	95.6	78.5	58.7	59.6	84.9	83.7	72.2	76.2
% with more getting done in the day	32.7	57.8	79.2	40.9	55.3	55.4	54.1	47.1	49.8	57.5	58.4	53.5
% using more for business than social calls	41.7	24.3	31.4	16.2	35.2	28.8	38.3	35.4	30.2	17.6	20.3	29
% using mobile phones for sending & receiving money	27.5	78.6	37.7	17.5	0.4	3	4.9	9.3	6	7.9	16.3	19
% using mobile phones for beeping	74.8	91.6	85.9	69.1	92.7	64.5	78	80.7	84.2	92.3	83.3	81.6
Average	52.2	63.1	55.9	43.3	53.6	43.9	46.2	43.6	49.6	48.8	48.1	49.8

Note Figures are users of a particular mechanism as a percentage of adult total mobile phone users in a country. Note also that the averages at the bottom of the table are just that: they are consistent with some people using all the mechanisms or a larger number using only one or two. Beeping is an activity which refers to deliberately missed calls. That is, an individual dials a number and then immediately cancels the call. Depending on the number of beeps it contains a message for the other party. It is classified as an economic mechanism because the goal is to save on costs
Source Research ICT Africa, survey data

M-PESA, as already noted, is a service developed by Vodafone and designed for emerging markets, where many people are still underserved by financial service providers. The first launch of M-PESA was in Kenya by Safaricom [in 2007] and one year later by Vodacom in Tanzania. Both Safaricom and Vodacom are part owned by the UK's Vodafone and hence have access to the M-PESA model (Camner et al. n.d.).[4]

Some of the reasons for the comparative success of M-PESA in Kenya are related to a receptive regulatory environment, the dominant status of Safaricom in the market and a strong demand for this firm's 'send money' function because of the high numbers of domestic migrants transferring funds back home (The Economist, 20 September 2012).[5] The absence of a comparable demand in Tanzania is one reason why adoption of the m-Pesa model in that country has lagged behind Kenya. Another reason has to do with the viability of the alternative to mobile phones in the two countries (see below and Chaps. 3 and 4 for a general version of this argument).

In Kenya the most popular methods previously used were asking a friend or family member to take the money by hand or to use a bus or a courier company. We know that these methods can be high risk and do result in losses. The high crime rate in Kenya and Nairobi in particular created a greater demand for a safe way of sending money compared to Tanzania where the risk of robbery is lower (Camner et al. n.d.).

For the other rows in Table 6.2, however, a more general explanation is needed for the superior performance of East African countries. What I will advance for this purpose is a variation of findings already published in the scant literature on the beneficiaries of mobile phone use. Waverman et al. (2005) for example,

> Find that mobile telephony has a positive and significant impact on growth and *this impact may be twice as large in developing countries compared to developed countries.* This result concurs with intuition. Developed countries by and large had fully articulated fixed-line networks in 1996… In developing countries, we find that the growth dividend is far larger because here mobile phones provide, by and large, the main communications networks; hence they supplant the information-gathering role of fixed-line systems (Waverman et al. 2005, p. 11).

In a case-study of Tanzania, South Africa and Egypt, Samuel et al. (2005) extend this logic to cover not just fixed line communications available to households, but also other means of contact such as ease and cost of travel and availability of well-functioning public pay-phones. They point for example to the travel time and cost savings achieved by a mobile phone call as opposed to travel as a means of communication. The savings were found to be higher in Tanzania than in South Africa because in the former 'roads are worse and public transport less extensive'. Similarly,

---

[4]It is worth noting that mobile money arose from within the Kenyan population before being formalized by Vodafone. The 'receptive regulatory environment' involved complex negotiations with the banking authorities who were initially opposed.

[5]Other articles on m-Pesa include Jack and Suri (2010) and IFC (2010).

In Tanzania, a strikingly high proportion of respondents (57 per cent) felt that a major impact from mobile phones was faster and improved communication. The proportion in South Africa mentioning this as an impact was substantially lower at 8 per cent. This probably reflects a greater presence and reliability of fixed-line phones in South Africa prior to the introduction of mobile phone services (Samuel et al. 2005, p. 49).

Generalizing these examples, the hypothesis is that the **more difficult and expensive it is to communicate by means other than mobile phones, the more (percentage) use will be made (*ceteris paribus*) of this technology**. And since it is assumed that the said difficulty varies inversely with country income, the prediction is that the **most intensive percentage use of mobile phones will occur in relatively poor countries**. (This assumption does not require a great leap of faith for it seems almost obvious that rail transport, paved roads and passenger cars vary directly with per capita income.)

The reasoning just advanced helps to explain the pattern observed in Table 6.2 for it is the five Eastern African countries that have the lowest incomes in the sample (That pattern is one which, among other things, includes countries from the region appearing in first place in all six rows). Of course, the explanatory framework I have adopted does not account for all entries in the table, especially those in which relatively rich countries are involved. For these cases, a country specific explanation may need to be involved (South Africa's second-placed ranking in the case of beeping may constitute one such example).

It is noteworthy in this regard to compare the results shown in Tables 6.2 and 6.3 with those of subscriptions for mobile phones in the same countries (shown in Table 6.4).

The interest lies in the almost diametrically opposite results of the two cases. On the one hand, it has already been shown that the poorest countries tended to make the most of mobile use whereas on the other hand these countries were heavily weighted against adoption of this technology. In this case it is the richest two countries—South Africa and Botswana—that occupy the top positions in adoption rates. Conversely, the East African countries which performed so well in rates of usage, fall into the bottom five places (with one exception) on this metric. Resolving this apparent paradox has to do as suggested above with the contrasting role of income in the two cases. As far as adoption goes, per capita income plays

Table 6.3 Leading countries according to economics related use mechanisms

	1st	2nd	3rd
% using a mobile phone to find work	Rwanda	Uganda	Cameroon
% with savings on time & travel cost	Ethiopia[a]	Kenya	Namibia
% with more getting done in the day	Tanzania	Botswana	Kenya
% using more for business than social calls	Uganda	Cameroon	Ethiopia
% using mobile for sending and receiving money	Kenya	Tanzania	Uganda
% using mobile for beeping (the caller dials but hangs up before the call is answered)	Ethiopia	S. Africa	Kenya

[a]Following numerous sources, Ethiopia is defined here as being part of Eastern Africa (e.g. Encyclopedia Britannica)

Table 6.4 Mobile cellular subscriptions by country, 2009 (in descending order)

Country	Cellular subscriptions (in percentages)
Botswana	96.1
South Africa	94.2
Ghana	63.4
Namibia	56.1
Kenya	48.7
Nigeria	47.2
Tanzania	39.9
Cameroon	37.9
Uganda	28.7
Rwanda	24.3
Ethiopia	4.9

Source World Bank, Little IT Book (2011)

a facilitative role, as it mostly does with the diffusion of durable goods (Rogers 1995; Stoneman 1983). But the benefits of use from mobile phones tend to be greater with lower income because this variable stands as a proxy for the difficulty of finding alternative ways to communicate (other that is than with mobiles).

I turn finally to compare the rows as opposed to the columns of Table 6.2. The most popular row in this table deals with beeping (at an average use rate of 81 %). In essence this phenomenon involves making a call to another mobile number and then hanging up before the call is answered in the hope that it will be returned by the other party in the future. Sometimes beeping is associated with a specific message such as when for example two rings mean 'I'm leaving now' (Donner 2007). In any event the cost of the original call is saved, an amount which presumably explains the popularity of the phenomenon in poor African countries (but also among the poor in relatively affluent sample countries such as South Africa). Together with mobile money, beeping is an interesting example of how important innovations can be made in and for African countries, which, to this extent, are not only reliant on information technology applications from the advanced countries.

Not far behind beeping is the second-highest mechanism (row) in Table 6.2 namely, the percentage with savings from mobile phones on time and travel costs. These savings have also been described as important in the study by Samuel et al. (2005) noted above. What they found was that,

> In the survey sample, 91 per cent of respondents in Tanzania called friends and relatives rather than travelling to see them. In South Africa, 77 per cent of mobile users called rather than visited.... Indeed, for many families surveyed the costs of travelling to see relatives would be prohibitive, especially in the poorest rural communities, and mobile therefore represented the only option of maintaining contact. ... The impacts were slightly larger for Tanzania, *where roads are worse and public transport less extensive.* The potential importance of mobile as a substitute for travel is easy to under-estimate. Of the communities surveyed in South Africa, only 4 out of 10 had a regular bus service to the nearest town and the typical round-trip cost was 15 Rand. In contrast, a typical pre-paid voice call cost R 5.... It is not surprising that so many respondents identify

Table 6.5   Leading countries according to health-related use mechanisms

Mechanism	1st	2nd	3rd
Setting alarm for medical appointment	Botswana	Uganda	Kenya
Setting alarm for taking medicine	Uganda	Rwanda	Kenya
Obtaining SMS reminder from clinic or doctor	Uganda	Rwanda	Kenya
Having mobile contact with health workers	Rwanda	Ghana	South Africa

> mobiles as a source of saving both time and travel cost. … one respondent in Mafia Island, Tanzania, said he was now able to keep in daily contact with his immediate family, who all lived in Dar es Salaam (Samuel et al. 2005, p. 49).

It should be clear that this particular explanation also relies on the view espoused earlier in the chapter that the use of mobile phones depends heavily on the availability and quality of the relevant infrastructure for alternative means of communication.

6.2.2   Health-Related Mechanisms

Tables 6.5 and 6.6 shows the use of mobile phones for four medical (m-health) applications. These data are also summarized, as in the previous section, with regard to the top three countries in each of the four mechanisms. Tables 6.5 and 6.6 are similar to their counterparts in the previous section in that Uganda and Kenya again play a significant role, though in this case Rwanda comes to the fore in a way that it did not do previously.

These results might well be explicable in terms of the relative amounts of foreign aid that are given for m-health applications (certainly Rwanda seems to receive a relatively substantial amount of aid for this purpose). Unfortunately, however, such comparative data are not readily available[6] and I shall try instead to explain why the three East African countries in question are especially amenable to this type of technology. Such a task is perhaps most easily accomplished in the case of Uganda. According to one observer for example,

> There is no doubt that Uganda has taken huge strides in mobile health. Whereas neighboring country, Kenya, has distinguished itself with mobile money schemes, Uganda has achieved much in widening access to health-related services based on mobile technologies. For example, the population of that country can easily check whether a doctor or clinic is licensed or not merely by sending an SMS. Whether by checking the diagnosis of HIV-infected babies or by educating people on sexual and reproductive matters …. Uganda is certainly showing itself to be East Africa's most prolific testing ground for m-health innovations (Mulupi 2012).[7]

---

[6]Available data deal with aid by sector rather than the use of IT by sector.

[7]For a fairly recent discussion of mobile phones and the health sector see Mulupi (2012). For an example of a successful Ugandan m-Health innovation see Luscombe (2012).

Table 6.6 Health-related mechanisms of mobile phones use

	Uganda	Kenya	Tanzania	Rwanda	Ethiopia	Ghana	Cameroon	Nigeria	Namibia	South-Africa	Botswana	Average of countries
% who set alarm for medical appointment	36.3	30	10.2	22.6	4.5	17.8	22.5	24.6	24.3	22.2	40.3	23.2
% who set an alarm for taking medicine	48.3	36.8	14.5	42.4	8.5	14	20.5	22	15.5	15.1	31.8	24.5
% who obtain SMS reminder from clinic or doctor	45.2	22	9.1	24.2	8.9	10.7	12.3	21.5	9.8	15.6	18.4	18
% having mobile contact with health care workers	59.6	0	62.5	83.8	67.4	78.8	18.1	69.9	10.5	75.5	22.4	49.9
Average	47.4	22.2	24.1	43.3	22.3	30.3	18.4	34.5	15	32.1	28.2	28.9

Note Figures are users of a particular mechanism as percentage of the total of adult users of mobile phones in a country
Source Research ICT Africa, survey data

What is special about Rwanda in this context is that it is rated second out of 123 countries by the World Economic Forum according to how highly ICT is prioritized by the government.[8] This distinction, accorded to one of the world's poorest countries, is partly a reflection of the support given to the technology at the highest levels of government. In particular, in that country, the president has set the pace with future-looking policies for e-health and mobile phones. The government's e-health plan is worth $32 million, and involves state leadership at the highest levels. Two components of the plan, rapid SMS alerts in emergencies and 'mUbuzima' monitoring tools for community health workers, are being introduced nationally (Mulupi 2012).

Kenya's frequent appearance in the table is due in part to the same institutions that were responsible for the M-Pesa programme in that country. Safaricom, for example, the owner of that successful programme, has also developed an m-health product ('Daktari 1525') which enables subscribers to contact doctors for advice at any time of the day or night. However, many of Kenya's mobile health applications have emerged from a new class of technological entrepreneurs who are often associated with 'iHub', one of Africa's most prominent IT incubators (BBC News, 19 July, 2012). On the user side, it is plausible to imagine that Kenyans are especially responsive to m-Health initiatives because so many of them are already accustomed—via M-PESA—to new applications of mobile phones.[9]

Note, finally, that in terms of rows rather than columns, the 'percentage having mobile contact with health care workers' is the most commonly used of the four mechanisms (with an average figure of almost 50 %). This is probably due to the fact that it is the most general of the mechanisms, including a wider variety of applications than the other three.

6.2.3 Social Capital Mechanisms of Use

According to many authors, IT has the potential to increase social capital. As such this technology will tend to increase existing patterns of social contact and civic engagement. The World Bank (2008) points specifically to 'bridging social capital which connects actors to resources, relationships and information beyond their immediate environment'. In these ways the participants will tend to gain from the increased social capital thus generated. Goodman notes that 'Social capital may be an even more important concept for developing countries than developed, as in many cases people in the former have less access to formalized structures of support such as the legal system or the financial system, and may rely on informal networks instead' (Goodman 2005, p. 54).

---

[8]For a description of these calculations see one of the Forum's annual reports.

[9]Note that there is a tendency in the m-Health literature to ignore the obvious predominance of voice communication by mobile, possibly because it seems less interesting in terms of technical innovations.

Following a suggestion by Granovetter, Goodman also distinguishes between strong links and weak links. The former 'are those between close friends and family, people who are regularly in contact and have a lot in common. Weak links are those between acquaintances or distant friends in irregular contact. Both types of links are crucial' (Goodman 2005, p. 57). One of the purposes of this section is to analyze the strength of the linkages shown in Table 6.7. First, however, let me deal with the relationships between the columns shown in that table and in the summary of best-performing countries contained in Table 6.8.

Table 6.8 shows that once again the East African group of countries performs better than a random influence would dictate. In particular, these countries take up 12 of the cells in the table compared to the 8 that would occur on the basis of chance alone (i.e. 5 out of the 11 countries in the sample). I continue to ascribe this performance to the difficulty of communicating—by other means than mobile phones—in the relatively low-income countries of the region. That said, however, the prominence of East Africa is lower here than in previous sections.

One question is whether the inclusion of Nigeria among the leading countries in Table 6.8 (twice first and once second) requires me to reconsider the role of low income in East Africa as an explanation of the previous patterns. In fact, it does not much alter that story because Nigeria has a per capita income just above the level of Kenya, the richest country in impoverished East Africa. The region's average income, that is to say, does not alter significantly after Nigeria is added to the group.

Another task is to explain what accounts for that country's excellent performance in some of the rows in Table 6.7. My suggestion—and it is only speculative—is that the forms of social capital most sought after by Nigerians—dealing with politics and religion—were strongly emphasized in the elections of 2011, a time at which the data were collected for the survey on which this chapter is based. There may, that is to say, be a causal connection between the dominant features of the election and the forms in which social capital were primarily sought at around the same time the data were collected. Even then, however, the direction of any such causality is not clear.

I turn finally to the rows in Table 6.7 and hope to establish some form of pattern between them. Unfortunately, however, only the highest position occupied by friends and family can be explained in terms of the distinction between strong and weak links described above. Numerous authors for example have alluded to the frequent and close relationships within this group and they also find what was shown in Table 6.7, namely, that it is a highly important empirical relationship relative to weaker links. But for the other entries shown in that table, it is impossible to say how frequent and intense were the contacts between parties. All that the data show are the percentage of people who seek to build up one or other of the contacts given in that table. Contacts with colleagues, for example, the next most important entry, could be weak or strong depending on the nature of the relationship.

Table 6.7 Social capital mechanisms of mobile phone use

	Uganda	Kenya	Tanzania	Rwanda	Ethiopia	Ghana	Cameroon	Nigeria	Namibia	South Africa	Botswana	Average
% using a mobile phone to mobilize the community or for political events	21.6	17.2	21.8	12.2	6.6	9.3	8	27.9	36.3	9.1	10.8	16.4
% using a mobile to increase their contact with those who share hobbies/sports	53.1	61.3	36.6	33.6	66	31.6	45.9	51.4	55.7	35.8	51.6	47.5
% using a mobile to increase their contact with those who share political views	32.3	34	21.1	26.2	16.8	19.9	19	42.7	26.1	15.7	27	25.5
% using a mobile to increase their contact with those who share religious beliefs	53.9	56.6	49.8	38.4	47.2	44.3	49.9	62.4	36.4	36.2	47.6	47.5
% using a mobile phone to increase their contact with family and friends	90.3	88.5	74.5	79.5	80.8	68.5	82	78.3	77.2	68.5	75	78.5
% using a mobile to increase their contact with colleagues	71.2	70.9	62.2	59.4	54.9	57	53.4	65.5	47.2	50.8	62.9	59.6
Average	53.7	54.8	44.3	41.6	41.6	45.4	38.4	43	54.7	46.5	45.8	45.8

Note Figures represent number of users of a particular mechanism as percentage of adult users of mobile phones in a country
Source Research ICT Africa, survey data

Table 6.8 Leading countries according to social capital mechanisms of use

Mechanism	1st	2nd	3rd
% using mobile phone to mobilize community etc.	Namibia	Nigeria	Tanzania
% using mobile phone to increase contact with those sharing hobbies	Ethiopia	Kenya	Namibia
% using mobile phone to increase contact with those sharing politics	Nigeria	Kenya	Uganda
% using mobile phone to increase contact with those sharing religion	Nigeria	Kenya	Uganda
% using mobile phone to increase contact with friends and family	Uganda	Kenya	Ethiopia
% using mobile phone to increase contact with colleagues	Uganda	Kenya	Botswana

6.2.4 Safety-Related Mechanisms of Use

The results for safety-related mechanisms of mobile phone use are shown in Tables 6.9 and 6.10. Beginning as usual with columns, Table 6.9 continues to highlight the prominent role played by countries from Eastern Africa, which, in the form of Ethiopia and Tanzania, make up two of the top four places.

For the next two cases, South Africa and Namibia, an explanation could take the form of a relative prevalence of violent crime.[10] After all, these countries belong to Southern Africa, a region that suffers from the highest rates of this type of crime in all of Africa (UNODC.org). The basic idea is that in places with high rates of violent crime, access to a mobile phone becomes more pressing (in terms of safety) than in countries where the rates are lower.

For the East African countries, however, this line of argument would seem to be largely irrelevant, since the crime rate there is neither unusually high nor exceptionally low. In these cases—Ethiopia and Tanzania—I fall back on a variant of the argument advanced in the three previous sections, namely, that income is a proxy for numerous other relevant variables. More specifically, I argue that in places with relatively high average income, access to a mobile phone becomes more pressing (in terms of safety) than in countries where the rates are lower. The point here being that income tends to stand as a proxy for other forms of communication than mobile phones, such as public transport, payphones, postal services, fixed line phones and telecentres. Where these are scarce or entirely non-existent there are few or no alternatives to mobile phones as a means of communicating with the authorities or others who can deliver protection (some mechanisms though are more about checking up on the safety of close friends or family, but in these cases, too, a central issue is about the lack of alternatives to the mobile phone).[11]

---

[10]For data see UNODC.org.

[11]The Economist points out that individuals are known to make reductions in even basic types of spending in order to buy time for their mobile phones ('Vital for the poor', November 10, 2012). See also 'Mobile Phone Use at the Kenyan Base of the Pyramid', iHub Research, 2012.

Table 6.9 Safety-related mechanisms of mobile phone use

	Uganda	Kenya	Tanzania	Rwanda	Ethiopia	Ghana	Cameroon	Nigeria	Namibia	South Africa	Botswana	Average
% who feel more secure from mobile phone	57.4	69.6	81	49.9	85.3	63.2	53.5	62.4	65.2	71.2	45.6	64
% using mobile phone for finding out about safety issues and to alert people	53.6	52.8	60	28.1	53	55.9	40.4	61.6	69.7	58	49.3	52.9
% using mobile phones to check on the safety of loved ones and to see where they are	81.5	79.3	85.7	66.2	94.6	85.5	67.6	76.9	87.8	92.1	88.6	82.3
Average = (cols)	64.2	67.2	75.6	48.1	77.6	68.2	53.8	67	74.2	73.8	61.2	66.4

Note Figures represent number of users of a mechanism as a percentage of total adult users of mobile phones in a country
Source Research ICT Africa, survey data

Table 6.10 Safety-related mechanisms of mobile phone use

Mechanism	1st	2nd	3rd
% who feel more secure from mobile phone	Ethiopia	Tanzania	South Africa
% using mobile phone for finding out about safety issues and to alert people	Namibia	Nigeria	Tanzania
% using mobile phones to check on the safety of loved ones to see where they are	Ethiopia	South Africa	Botswana

I turn now to examine the rows of Table 6.9. Noting that safety is an important component of basic human needs (Maslow 1943) it is worthwhile to ask how much the mobile phone contributes to it. Such a question has unfortunately not been widely posed in the literature on the welfare impact of mobile phones. Yet, as Table 6.9 suggests, the relationship between mobiles and safety appears to be a powerful one. Indeed, as many as 82 % of the sample use this technology as a way of checking on the safety of close friends and family. In addition 64 % of the sample feel more secure as a result of having a mobile phone. Because of figures such as these, the appendix to this chapter focuses in more detail on the relationship between mobile phones and safety.

6.3 Cross-Tabular Analysis

So far my analysis has been confined to individual tables grouped according to economics, social capital and so on. In this section I seek to examine the results at a cross-tabular level, as if the groupings no longer mattered. Which mechanism, for example, is the strongest (on average) across all of the four tables presented above? And which countries (again on average) achieve the highest scores across all these tables? The (partial) answer to both questions is contained in Table 6.11. The left-hand column lists the highest-rated mechanism across all four tables, whereas the column on the right shows the highest ranked countries across all the mechanisms that were described above.

The most striking comparison between the two columns lies in the variety of groups from which the mechanisms and countries are drawn. In terms of the first column, the entries are drawn from the economic, social capital and safety

Table 6.11 Cross-tabular results (rows and columns)

Mechanism	Country
% using a mobile phone to check on the safety of loved ones (safety) 82.3	Ethiopia 77.6 (safety)
% beeping (economics) 81.6	Tanzania 75.6 (safety)
% using mobile phone to increase their contact with friends and family (social capital) 78.5	Namibia 74.2 (safety)
% with savings on time and travel cost (economics) 76.2	South Africa 73.8 (safety)

mechanisms, in contrast to the second column where all the entries are drawn only from the last-mentioned group (safety). Taken together, the two columns indicate that safety is an important (and in my view neglected) aspect of the relationship between mobile phones and individual welfare.

The need for safety is perhaps to be expected in Africa where crime rates are on average higher than in many other parts of the developing world. The two cases from Southern Africa, South Africa and Namibia, are particularly clear examples of this point. But it is also true that safety is a fundamental need which comes just after physiological needs in Maslow's (1943) well-known hierarchy. Researchers would do well to bear this in mind when examining the welfare effect of mobiles.

Of the economics-related mechanisms in the left-hand column of Table 6.11, two stand out, namely, beeping and saving time and travel costs. The former is especially important in places where income is low and the need to save on costs of communication is correspondingly high. That is why it tends to be relatively poor East African countries which dominate these categories (though Nigeria has a per capita income just above the richest of those countries, Kenya). I have already referred to research which shows that savings in time and travel costs are substantial for two African countries and the importance of beeping has been well established in the literature. In short, the presence of these two economics-related mechanisms in Table 6.11 is no great surprise and serves only to confirm on a larger scale what others have already found.

6.4 Conclusions

(1) This chapter has sought to analyze an extensive large-scale survey of mobile phone use across eleven African countries, ranging from the relatively rich (such as South Africa) to the relatively poor (such as Ethiopia). These data comprise one of the first comprehensive studies of mobile phone use in developing countries.

(2) Since use is a better indicator of welfare than penetration, the results give as comprehensive a view of the micro-impact of mobile phone use as is now available for Africa and indeed developing countries as a whole (though as noted there are a few relatively limited case studies on the topic).

(3) The most important general finding is that East Africa is a prominent region in all four mechanisms of use I have identified, namely, economics, social capital, health and safety.

(4) Part of the explanation for this finding is country specific: the role of mobile money in Kenya, m-Health in Uganda and the role of the president in promoting mobile phones in Rwanda's health sector. Another part however has to do with the fact that East African countries are the poorest in the sample.

(5) This is important according to the following line of argument: the more difficult and expensive it is to communicate by means other than mobile phones, the more use will tend to be made of this technology. If it is assumed that

the said difficulty varies inversely with per capita income, the prediction is that mobile phones will be most intensively used in relatively poor countries (which are especially lacking in a technological infrastructure in the form of paved roads, post offices, taxis, public phones, rail transport and so on). When use is taken into account, therefore, the degree of inequality between rich and poor countries will tend to diminish and the digital divide between these countries will become smaller. These predictions are examined and tested in Chap. 7. If these are confirmed, it will further underline the ubiquity of a pro-poor influence at work in the analysis of the impact of mobile phones in developing countries.

(6) The most important mechanisms in descending order are: using a mobile phone to check on the safety of family and close friends, 'beeping', using a mobile phone to increase contact with friends and family, and saving on time and travel costs.

(7) Safety is not widely discussed in terms of the benefits of mobile phones in developing countries. Yet, it is often described as a fundamental human need, coming second only to basic physical desires such as hunger and thirst. And the data show strong support for the connection between mobile phones and safety. For example, 82 % of the sample use this technology as a way of checking on the safety of close friends and family. Some part of that figure will comprise the bottom 40 % of the population. For further analysis of mobile phones and safety see the appendix to this chapter.

(8) There remain numerous anomalies to explain. Why, for example, does South Africa perform so well in some tables and so poorly in others? Do the results suggested for eleven African countries hold when many more countries from that region are included?

Appendix 1: Mobile Phones and Safety in Developing Countries: Further Evidence

The pursuit of safety seldom appears in the existing literature on the use of mobile phones in developing countries; nor is it to be found among the most important uses of this technology in developed countries. However according to detailed evidence cited above for 11 African countries and a sample of Asian nations, it turns out that this need appears (on average) at the very top of a list of uses of mobile phones in the countries concerned. The survey data also enable countries to be ranked according to the safety mechanism in question. The purpose of this short appendix is to document, and where possible explain these novel findings. My main thesis is that the observed desire for safety via mobile phone use is strongly driven by the extent of crime in a country, which, in turn, is closely related to existing levels of poverty and inequality. By way of empirical support for this argument I rely heavily on relatively recent cross-country survey data compiled by the United Nations Office on Drugs and Crime (UNODC 2011). First, though, I deal with other sources of information that link mobiles to safety.

Appendix 1.1: Non-conflict Situations and Safety

Using the context of fishing in Kerala, Sreekumar (2011) argues that there is a connection between mobile phone usage and safety even in non-conflict situations: "The possibility of actualizing the collectivist logic in a community's appropriation of a new technology has been overlooked in cultural studies on mobile phones" (Sreekumar 2011, p. 175). Accordingly, he "examines how a subaltern fishing community in Kerala, India, has appropriated the cell phone in ways that are in keeping with its traditional collectivist ethos. Within this context of collectivist appropriation, the cell phone has helped enhance their working and living conditions" (2011, p. 173).

More specifically, Sreekumar (2011) studies how mobile phones contribute to the safety of fishermen during extreme weather conditions and emergencies. The importance of mobile phones in these conditions is reflected in the seriousness with which the safety of the technology is ensured while at sea. Many fishermen, for example, described that the phone is kept safely in a polythene sheet inside a box that is tied to the bottom of the engine. His main finding is that the "primary narrative that defines the discourse of fish workers when they mention mobile phones has to do with its utility during emergency situations" (2011, p. 178).

On rare occasions, however, even fishing can turn into a conflict situation. In West Africa, for example, pirates have sometimes resorted to violence against local fishermen who, as a result, became fearful of taking mobiles to the local seas. The problem was dealt with by distributing mobile phones equipped with GPS-enabled cameras, which were then used to report the presence of pirates to government and other institutions that are charged with following-up on the illegal fishing vessels.

Appendix 1.2: Civil Wars and Post-conflict Situations

The role of mobile phones in conflict and post-conflict situations is rather similar in that diminished safety is a hallmark of both these circumstances and the need to check on the safety of friends and families is ubiquitous. Best (2011), for example, in speaking of post-conflict Liberia points to a "consistent prevalence of security and emergency uses" (p. 24). Moreover, Monrovian data suggest "a distinct factor emphasizing security, while rural data revealed security as an item of consensus" (Best 2011, p. 24). It was clear that the safety and security of self, of loved ones, and of personal property is also still a concern in Liberia.

Table A.1 Top use-mechanisms (cross-country average)

Mechanism	%
% using a mobile phone to check on the safety of loved-ones	82.3
% beeping[a]	81.6
% using mobile phones to increase their contact with friends and family	78.5
% with savings on time and travel cost	76.2

[a]*Note* Beeping is a process of calling without picking up to get someone to call back or to convey a message through a set number
Source Research ICT Africa (2011)

One of the most interesting ways in which mobile technology contributes to safety was provided by an operator in Liberia. The operator recounted that after his company had announced its intention to remove free calling during the night hours, customers complained about their needs during emergency situations. Another operator suggested that for many users night time safety was more important to them than conserving battery power (Best 2011).

Appendix 1.3: Survey Evidence

Consider from Tables A.1 and A.2 which are about, respectively, the leading uses of mobile phones (averaging across countries) and the countries that rank highest according to the top use mechanism—the one, that is, which is concerned with the safety of family members. Table A.1 shows that no fewer than 82.3 % of mobile users employ the technology for this purpose.[12] Evidently the ability to check on the safety of such persons is even more important than the familiar benefits of saving time and travel costs via mobile phones in environments where alternative modes of communication are meager, if not entirely absent. It is important to note here that Africa is not the only region in which the primary importance of safety emerges from survey research. In particular, research conducted in a sample of Asian countries, among those at the BoP, found that having a mobile phone provides a key benefit in terms of security (inter alia), and that is often the motivation for them getting connected in the first place (Lirneasia, n.d.). Indeed, out of a list of possible uses of mobile phones 'the ability to contact others in an emergency is the highest perceived impact' (Lirneasia, nd). This result is especially important because it deals with survey respondents at the BoP and it is the impact of mobile phones on such individuals that is my ultimate concern. That concern, moreover, may need to embrace a gender component if in fact these technologies 'provide females with a sense of security that is considered less necessary for men' (Castells et al. 2007, p. 45).

---

[12]Several other safety-related mechanisms are also covered by the survey but I do not deal with them here.

Table A.2 Countries ranked according to safety use-mechanisms (%)

Country	%[a]	Rank
Ethiopia	94.6	1
South Africa	92.1	2
Botswana	88.6	3
Namibia	87.8	4
Tanzania	85.7	5
Ghana	85.5	6
Uganda	81.5	7
Kenya	79.3	8
Nigeria	76.9	9
Cameroon	67.6	10
Rwanda	66.2	11

[a]This refers to the percentage of people who agreed with the statement that 'I use my mobile phone to check on the safety of my loved ones, and to see where they are'

One key question for analysis therefore is why safety-related use is so important to so many respondents in the countries concerned.

Further analysis is suggested by the ranking of countries on the use mechanism in question, as shown in Table A.2. Consider in particular that the four countries listed at the top of this table comprise an unlikely mixture. On the one hand, Ethiopia is the poorest country in the sample and the three others constitute the richest. Clearly, no simple explanation based on income (or variables that are closely related to it) will suffice. I cannot provide a fully satisfactory alternative but suggest that it has something important to do with the relationship between crime, poverty and inequality.

Appendix 1.4: Crime, Poverty and Inequality

Cross-country level data provided by UNODC (2011), show first that at this level there is a clear correlation between (violent) crime and per capita income. 'Higher levels of homicide are associated with low human and economic development. The largest share of homicides occur in countries with low levels of human development' (UNODC 2011, p. 10). Furthermore, 'Inequality is also a driver of high levels of homicide. Homicide rates plotted against the Gini Index, … show that at global level countries with large income disparities … have a homicide rate almost four times higher than more equal societies' (UNODC 2011, p. 30). The correlative evidence just described also holds for Africa, in whose poorer regions one third of the world's homicides occur.

Southern Africa suffers from particularly acute levels of violent crime (noting that South Africa, Botswana and Namibia all come from this region). Indeed, in

2010 six of the world's most unequal societies were located in Southern Africa. (World Bank, indicators of Gini coefficients 2013).

Poverty alone exerts an influence over violent crime partly because the evidence suggests that young people join gangs and rebel groups because of unemployment. Yet, it is important to recognize that the relationship between crime and poverty is a complex and multifaceted one, as the favelas of Latin America seem to suggest (Jones and Rodgers 2011). In those locations the prevalence of drugs, gangs and gambling interact to create and maintain desperately high levels of poverty. To these effects should be added the observation that Southern African countries have long suffered from the problem of 'grafted capitalism' (OSISA, n.d.). 'During colonization...., the capitalist sector of the economy was grafted onto a pre-capitalist form of production in a distorted manner. This kind of capitalism did not transform the economy as a whole but only a small enclave sector thus failing to produce growth and development. The small formal sector was totally dependent on external factors such as markets in, and capital from Europe' (p. 2). More specifically, the enclave exists alongside an underdeveloped peasant-based subsistence rural economy and an urban informal economy; hence the formal sector accounts for less than 20 % of the labour force (OSISA, n.d.). Because of minimum wages and union activities, wages in the formal sector are usually higher in comparison to what the unfettered market would dictate. The situation with respect to inequality in these countries was poor even at the outset of colonialism. Not much has changed structurally since then. Currently, Southern Africa's extractive industries have further fuelled inequality and poverty. They have deepened enclave developments as the extractive zones became the center of government and private sector attention. Thus, 'while oil, copper, gold, diamonds, chrome ... are in plentiful supply in the SADC (Southern African Development Community) region, unemployment is increasing, poverty is deepening and inequality between and within countries is widening' (Jauch 2011, p. 1).

Much of the persistence of this type of situation has to do with political economy, namely the alliance between the government and the mineral enclave part of the economy. There is an overlap too between the interests of foreign and local capital, neither of which has an interest in changing the status quo (Ross 1999). Thus despite the rhetoric in the post-apartheid era in South Africa, the degree of inequality remains at an exceptionally high level.

According to strain theory, by sociologists, when confronted with the relative success of others around them, unsuccessful individuals feel frustrated and the higher the strain (due also to say ethnic fractionalization), the greater the incentive to commit crime (with all else being equal). The 'tunnel effect' is also relevant here. In the short run, welfare is likely to increase as the prospects for moving up go up; if however the situation persists welfare is likely to decrease as frustration sets in. The tunnel analogy can be profitably used to explain the tolerance for inequality as poor countries grow. Suppose in particular that "an individual's welfare at any point of time depends on both his present as well as expected future level of contentment ... Although the individual generally has good information about present income, his information about future income may be far more limited' (Ray 1998, p. 200).

Consider next an improved situation on the part of some of those next to him. How the individual responds to this improvement will depend on his view about the probable effects of it on his own welfare. He may for example believe that the improved social or economic position of others augurs well for him in the fore-seeable future. In that case his welfare will not be diminished by what has happened to others. Indeed, he may even experience an increase in his own sense of wellbeing.

Hirschman and Rothschild described such an increase in an individual's utility (and hence a tolerance of greater inequality) resulting from an improvement in others' economic status as the *'tunnel effect'* (Ray 1998, p. 200). The welfare situation would be reversed however if the improvement in the economic situation of others is judged as no indication of the individual's prospects. For example, society may be sharply divided on racial, cultural and other dimensions. The improved welfare of someone from another racial or ethnic group would then indicate little about the individual's future situation.

Thus, in relatively heterogeneous societies there will tend to be a relatively low tolerance for inequality and conversely in more homogeneous groups the reverse will apply. Suffering as many of them do from pronounced ethnic fractionalization and other divisions, countries south of the Sahara will tend on average to fall into the former category. South Africa is an especially clear example of this because of its acute divisions along ethnic lines.

In the specific context of the three Southern African countries under consideration—South Africa, Botswana and Namibia—the economic concept of resource abundance needs to be added, because it has been argued that dependence on say, minerals, creates income inequality for two reasons: In the first place resource extractive industries tend to be of an 'enclave' type, exhibiting few linkages into the local economy. Second, where public expenditure is biased in favour of the formal sector, it may worsen inequality (ODI 2006).

Appendix 1.5: Conclusions

My analysis has been provoked by the absence in the literature of evidence about and discussion of the relationship between mobile phone usage and safety. Recent survey findings for Sub-Saharan Africa and Asia suggest that the most important way in which this technology is used on average has to do with safety. Further impetus to investigate these findings comes from the pattern of countries that score highest according to this particular use mechanism, which varies from the very poorest to the very richest. Explanation of these outcomes, I have suggested, has to do with the relationships between poverty, inequality and crime. It was facilitated in each case by recent data concerning crime and development, but also by some intuitive theorizing as well.

Several lines of research are suggested by the main findings. As far as the relationship between resource abundance and inequality is concerned, the question may be less about whether causality runs from resource abundance to inequality and more about whether an abundance of say oil leads to 'successful' or 'unsuccessful' developmental outcomes. There is often much that can be learnt by countries in the latter group from those in the former.

Then, in a similar vein, there are innovative projects to alleviate the safety concerns of those caught in conflict/post-conflict situations. One of them is the use of Peace Huts that are linked by mobile phones to the police in Liberia (UN Women 2012). The idea is that the phones are distributed by the Liberian National Police 'to women participating in the country's peace hut initiative to help prevent crimes and violence against women. ... In addition to the cell-phone distribution, a free hotline to the police was established with private sector support to facilitate calls' (UN Women 2012, p. 1).

Finally, I have shown that a relationship between mobile phones and safety occurs in diverse geographical situations. These range from non-violent circumstances such as fishing to civil wars and post-conflict societies. Civil wars are especially prominent in Sub-Saharan Africa due mainly to poor economic conditions, low income, previous low growth and high dependence on natural resources (Hoeffler 2008). In these and other circumstances where violence is prevalent, it will be worth investigating whether safety is especially important to particular groups, most notably women. And if so, the question is how this need can best be met by policy towards mobile phones (including through for example the peace hut initiative mentioned in the previous paragraph).

References

Camner G, Sjöblom E, Pulver C (nd) What makes a successful mobile money implementation? Learning from M-PESA in Kenya and Tanzania. GSMA. Available at http://www.gsma.com/mobilefordevelopment/—/mpesa. Accessed 27 Oct 2013

Dewan S, Ganley D, Kraemer K (2004) Across the digital divide: a cross-country analysis of the determinants of IT. Working Paper, Graduate School of Management, University of California, Irvine

Donner J (2007) The rules of beeping: exchanging messages via intentional "missed calls" on mobile phones. J Comp Med Comm 13(1):1–22

Goodman J (2005) Linking mobile phone ownership and use to social capital in rural South Africa and Tanzania, The Vodafone Policy Paper Series, 3

Hellström J (2010) The innovative use of mobile applications in East Africa. 12, SIDA Review

International Finance Corporation (IFC) (2010) M-Money channel distribution case-Tanzania. Available at www.ifc.org/—/Tool%286.8%28Stud. Accessed 29 July 2012

Jack W, Suri T (2010) The economics of M-PESA: an update. Available at http://www.mit.edu/~tavneet, M-PESA.pdf. Accessed 22 Dec 2012

James J, Versteeg M (2007) Mobile phones in Africa: how much do we really know? Social Indic Res 84(1):117–126

James J (2009) Sharing mechanisms for information technology in developing countries: social capital and quality of life, Soc Indic Res 94(1)

Luscombe B (2012) TIME's mobile tech issue: tracking disease, one text at a time. TIME Magazine, August 15

Maslow A (1943) A theory of human motivation. Psych Rev 50:370–396

Mulupi D (2012) Uganda's fast rise to m-Health Hub. Available at Ventures Africa.mht. Accessed 3 Jan 2013

Qiang C, Iamamichi H, Hausman V, Altman D (2011) Mobile applications for the health sector, World Bank

Rogers E (1995) Diffusion of innovations. The Free Press, New York

Samuel J, Shah N, Hadingham W (2005) Mobile communications in South Africa, Tanzania and Egypt: results from community and business surveys. The Vodafone Policy Paper Series, 2

Stoneman P (1983) The economic analysis of technological change. Oxford University Press, Oxford

Waverman L, Meschi M, Fuss M (2005) The impact of telecoms on economic growth in developing countries. The Vodafone Policy Paper Series, 3

World Bank (2008) Social capital and information technology. Available at http://web.worldbank.org/WBSITR/EXTERNAL/TOPICS/EXTSOCIALDEVELOPMENT. Accessed 9 Feb 2011

Appendix References

Best M (2011) Mobile phones in conflict-stressed environments: macro, meso and microanalysis'. In Poblet M (ed) Mobile technologies for conflict management: online dispute resolution, governance, participation, law, governance and technology, vol. 2. Springer, Berlin

Castells M, Fernandez-Ardevol M, Qui J, Sey A (2007) Mobile communication & society: a global perspective. MIT Press, Cambridge MA

Hoeffler A (2008) Dealing with the consequences of violent conflicts in Africa. Background paper for the African Development Bank Report

Jauch H (2011) Time to turn the tide. Friedrich Ebert Stiftung. Available at http://library.fes.de/pdf-files/iez/08221.pdf. Accessed 16 April 2012

Jones G, Rodgers D (2011) The World Bank's World Development Report 2011 on conflict, security and development. J Int Dev 23(7):985–995

ODI (2006) Meeting the challenge of the resource curse, London

OSISA (nd) Inequality in Southern Africa: options for redress. Policy Brief. Johannesburg

Ray D (1998) Development economics. Princeton University Press, Princeton

Research ICT Africa (2011) Mobile data. Cape Town

Ross M (1999) The political economy of the resource curse. World Polit 51:297–322

Sreekumar T (2011) Mobile phones and the cultural ecology of fishing. The Info Soc, 27(3)

UN Women (2012) From conflict resolution to prevention: connecting Peace Huts to the policy in Liberia, United Nations Entity for Gender Equality and the Empowerment of Women, New York

UNODC (2011) Global study on homicide, Vienna

Chapter 7
Mobile Phone Use in Africa: Implications for Inequality and the Digital Divide

Abstract There is growing recognition that the welfare effects of mobile phones—as with other products—need to be based not only on the adoption but also the use of the technology (Sen 1985). What is increasingly being acknowledged is that adoption provides only a part of the true welfare effect because information about the purchase of goods in themselves tells us nothing about how they are actually used (In the extreme case for instance technologies may not be used at all, as when, for example, a bicycle is given to a crippled person) (Sen 1985). The bicycle thus has a positive effect on the GDP but confers no actual utility. In theoretical terms what is being proposed here is a movement away from traditional theory where welfare occurs at the point of purchase to a theory (such as Sen's functionings approach) which explicitly examines the process after a good is purchased. Methodologically, this transition is effected by means of a detailed data-set for 11 African countries, which describes how mobile phones are used for a variety of mechanisms involving economics, health, social capital and safety (The survey is conducted by the same institution that is indicated in the previous chapter.). These data were also employed in Chap. 6 to analyze use patterns among those countries.

Keywords Use-scores · Adoption rates · East Africa · Communication · Health · Safety

I begin with a statement of the major hypotheses of this chapter.

7.1 Hypotheses

The main theoretical basis of this brief chapter is the view that the more difficult and expensive it is to communicate by means other than mobile phones, the more (percentage) use will be made of this technology. This assumption, as noted in Chap. 6, does not require any great leap of faith since infrastructure in its broadest sense, involving rail and road transport, taxis and buses, paved roads and so on, tends to vary directly with per capita country income. For example, more use would tend to be made of mobile phones for communicating with friends and relatives in a country, lacking relatively, in roads and public transport.[1] Similarly, the rich have little or no use for mobile money since they have access to an entire formal banking system. Or again, some mobile health applications are less relevant to those who have easy access to doctors and hospitals (with the use, say, of cars and adequate public transport).

If examples such as these can be generalized and adoption varies directly with per capita income, then there should be a negative relationship between intensity of use (as measured below) and income per head of the countries in the sample. And if this is indeed the case, then the inclusion of the use variable offsets the inequality in adoption rates of mobile phones in the countries concerned. As far as the digital divide is concerned, the familiar ratio of mobile phones in rich compared to poor countries is replaced by the ratio of mobile phones multiplied by the overall use intensity for the two groups of countries. Because use is thought to be inversely related to per capita income, the digital divide will then fall with the inclusion of this variable (where the divide is defined in this context as rich vs. poor African countries in the sample).

Though the logic to this argument may seem very similar to the welfare analysis of previous chapters, there is an important conceptual distinction between them. In those chapters, that is, welfare was closely related to what tools of communication and sources of information were available to users *other* than a mobile phone. The lesser, or the more expensive, are those alternatives, the greater tend to be the benefits from the use of mobile phones. And given the usual lack of other such options (due often to a weak infrastructure) among poor individuals and countries, that amount would tend to be higher than it is among the better-off, who, might, for example, have fixed-line phones as an alternative to the mobile technology (whereas the poor may have to walk long distances to communicate with others). This is basically the reason why the consumer surplus was found to be higher among the poor than the non-poor in the case of Grameen Bank's VPP initiative in Bangladesh and the African examples discussed by Samuel et al. (2005). In this chapter, by contrast, the benefits derive from a different source, namely, the intensity with which the technology is used. The higher this is, the greater will welfare tend to be. Another difference is that I deal here with countries rather than the individuals of most previous chapters.

7.2 Testing the Hypotheses—Calculating Use Scores

The mobile phone use mechanisms listed in the survey are divided into four groups as shown in Table 7.1.

---

[1]For more examples see the study by Samuel et al. (2005).

Table 7.1 Groups of use mechanisms

Economics related	Health related	Social capital related	Safety related
• % using mobile to find work • % with saving on time and travel cost • %with more getting done in the day • % using more for business than social calls • % using mobiles for sending and receiving money • % using mobile phones for beeping	• % who set alarm for medical treatments • % who set alarm for taking medicine • % who obtain SMS reminder from hospital or doctor • % having contact with health-care workers	• % using a mobile phone to mobilize the community or for political events • % using mobile to increase contact with those who share hobbies/sports • % using mobile to increase contact with those who share political views • % using mobile to increase contact with those who share religious beliefs • % using mobile to increase contact with family and friends • % using mobile to increase contact with colleagues	• % who feel more secure from mobile phone • % using mobile phone for finding out about safety issues and to alert people • % using mobile to check on the safety of loved ones and to see where they are

Source Research ICT Africa

Table 7.2 Use scores by country

Country	Score
Uganda	21
Kenya	17
Ethiopia	17
Tanzania	9
Nigeria	8
Botswana	7
South Africa	6
Rwanda	5
Namibia	4
Cameroon	3
Ghana	2

Source Research ICT Africa, own calculations

For each such mechanism I have assigned a score of 3 for a first-place ranking, 2 for a second and 1 for a third. Aggregating over these scores I arrive at the overall result shown in Table 7.2.

Note that for some of the reasons given in Chap. 6, the top four countries are all from East Africa. For example, Uganda's success in mobile health use was attributed to that country's focus on mobile applications in the sector.

7.3 Testing the Hypotheses—Results

Apart from the use scores thus calculated, two other pieces of data are required to test the hypotheses mentioned above. One of them is adoption rates of mobile phones by country and the other is per capita income levels of each country. All three sets of data are presented in Table 7.3.

Note first that adoption closely correlates (0.87) with per capita income levels. This means that if the latter are unequal, as they certainly are in our case, so too will be the pattern of adoption (the income levels vary from rich African countries such as Botswana and South Africa to the very poor such as Ethiopia and Uganda). The correlation between use scores and adoption, however, is negative and moderately strong (0.45). Evidently, it tends to be the poorer countries that use mobile phones most intensively, thus offsetting the inequality observed with adoption alone (Indeed, the top four countries on use scores all belong to the category of very poor; see Table 7.2).

Income differences alone, however, do not tell the whole story as is obvious from the size of the correlation coefficient with use rates. For, there are several occasions where the high-income countries such as South Africa, Namibia and Botswana appear in the top three countries in use-intensity. In such instances other influences than income need to be invoked. For example, the fact that Southern African countries tend to be more affected with violent crime means that they are prone to use mobile phones more intensively for safety purposes (see the discussion in the appendix to Chap. 6).

With respect to the digital divide as well, the situation changes substantially when mobile use is added to adoption alone. To show this I group the sample into two categories: the richest 5 countries (South Africa, Botswana, Namibia, Ghana and Nigeria) and the poorest five (Ethiopia, Uganda, Tanzania, Rwanda and Kenya). I then took the average adoption rates for the two groups and divided the

Table 7.3 The basic data

Country	Use score	Adoption rate (%)	Per capita income ($)
Uganda	21	28.7	1.3
Kenya	17	48.7	1.8
Ethiopia	17	4.9	1.1
Tanzania	9	39.9	1.5
Cameroon	3	37.9	2.3
Botswana	7	96.1	16.2
South Africa	6	94.2	11.1
Namibia	4	56.1	7.5
Nigeria	8	47.2	2.6
Rwanda	5	24.3	1.3
Ghana	2	63.4	3.1

Source Column 1 from Table 7.2; Research ICT Africa, CIA World Fact book

'rich' African figure by the 'poor' (i.e. 71.4 divided by 29.3). The result, 2.4, is a measure of the 'African' digital divide when only adoption counts. Mobile phone use is then introduced by multiplying each adoption figure by the relevant use percentage. When this is done, the divide disappears entirely because, once again, the pro-poor character of phone use acts as an offset to the relatively inegalitarian nature of adoption alone.[2]

Note though that on the basis of the argument advanced in Chap. 4 about leapfrogging, mobile phones also have an egalitarian impact on adoption, by making this technology especially attractive to low-income individuals and countries (recall, in particular, the discussion there about leapfrogging characteristics). Even so, however, there remain poor countries where adoption countries do lag well behind the average for developing countries, which itself lies well below the average for developed countries. In 2013, for example, the average number of subscribers per 100 people in low-income countries was 55 as compared with 121 in high-income areas (World Bank, indicators). And below the average for the former countries, are extreme cases such as Niger and Myanmar with subscription rates of only 39 and 13 respectively.

7.4 Conclusions

Use intensity of mobile phones seems, tentatively, to have an equalizing influence on inequality between countries as suggested in Chap. 5. For this reason effective use needs to be promoted as widely as possible in poor developing countries. Much more needs to be known, however, about the non-income influences on this variable. More needs to be known too about the generalizability of these findings, which are based on only a small number of countries. For, if they do apply to developing countries as a whole, then the very notion of a global digital divide gets called into question. The more so since, as I have explained in previous chapters, it is the use rather than the adoption of new technology that is the more appropriate measure of welfare.

References

James J (2013) Digital interactions in developing countries. Routledge, London
Samuel J, Shah N, Hadingham W (2005) Mobile communications in South Africa, Tanzania and Egypt: results from community and business surveys. The Vodafone Policy Paper Series, no. 2
Sen A (1985) Commodities and capabilities. North-Holland, Amsterdam

---

[2]For other criticisms of the digital divide, see James (2013).